THE DEATH OF AIR INDIA FLIGHT 182

Dedication

This book is dedicated to the 329 victims of the sabotage of Air India Flight 182 and the two victims of the Narita bomb, as well as their families. Also to the Royal Canadian Mounted Police for their extraordinary work, members of the Canadian Security Intelligence Service, who must remain in the shadows, and investigators in the United Kingdom, Ireland, Japan, the USA and India. Finally, to my wife and children for all their help.

THE DEATH OF AIR INDIA FLIGHT 182

Salim Jiwa

Editorial Assistant: Don Hauka

A STAR BOOK
published by
the Paperback Division of
W.H. ALLEN & Co. PLC

A Star Book
Published in 1986
by the Paperback Division of
W.H. Allen & Co. PLC
44 Hill Street, London W1X 8LB

Copyright © Salim Jiwa 1986

Printed and bound in Great Britain by
Anchor Brendon Ltd., Tiptree, Essex.

ISBN 0 352 31952 6

PART ONE

THE DISASTER

1

A Passage to India

Vancouver, 22 June 1985, 7:00 a.m.

Sukhwinder Uppal, her two children and her friend Pradeep Sidhu bowed respectfully as they entered the domed Sikh temple in Vancouver. The morning prayers were just coming to an end as the four entered the main prayer hall. They walked towards the priest, who sat on the red carpet in front of the small podium with his legs crossed under him, and sat down before him. With her head bowed, her voice barely higher than a whisper, Sukhwinder asked for a prayer for guidance from the holy man, a portent for her trip halfway around the world to her homeland, India.

The turbaned priest picked up the *Guru Granth Sahib* – the Sikh Holy Book. A smile played beneath his salt-and-pepper beard. He had offered many such prayers in his years of service to followers of the monotheistic faith, for no devout Sikh ever undertakes a major venture without a prayer and a *Hukam Nama* – a divine order – for this is required by an ancient tradition. The priest opened a page at random and started to read the first verse on which his eyes fell, his voice rising and falling rhythmically with the cadence of the words.

'Man does not decide who lives on this earth and who goes away. The decision of life and death is made by God,' read the learned man of God, in the tongue in which the book was written – Punjabi. As Sukhwinder sat with her head bowed, contemplating the holy pronouncement, her friend Pradeep looked up sharply at the mention of death. The priest's words had made her heart skip a beat. Sukhwinder Uppal was like a sister to her. She knew well the life of hardship the gentle 37-year-old widow had endured, the trail of disasters that had followed her and robbed her of the joys of life. The verse was a reminder to both women that we are all creatures of destiny. The passage chosen by chance from the *Guru Granth Sahib* is accepted by the faithful as the divine decision on the task at hand.

The two women and the children stood up and bowed, retreating from the priest and the podium. As they walked out of the temple towards their car, Pradeep's thoughts were still about the *Hukam Nama*. The mention of death had taken her back to the terrible day in May 1975 when Sukhwinder, two months pregnant, was left a helpless widow by a car accident. The crash was a double blow for the tight-knit family. It had claimed both her husband Harbhajan, who was looking forward to the birth of the new baby and hoping he would have a son, and Sukhwinder's brother Dilbar Sidhu. With her only daughter Parminder just learning to walk and the baby on the way, Sukhwinder was devastated. Her entire family went into shock. The young widow did not smile again until the day in December when she gave birth to her son, Kuldip.

Living in the basement of her father's house in South Vancouver, Sukhwinder got on with the task of raising two little children on her own. She got a job with a drycleaning firm and led a quiet, earnest life, her intense faith helping her to carry on. But once in a while, memories of her husband would overwhelm her, and she would retreat into a quiet corner of the house and cry silently long into the night.

Her dream through these dreary years was to travel to her birthplace with her children. Slowly the dream was taking shape. Pradeep, a travel agent with Gaba Travel in Vancouver, had done everything she could to make the trip easier for the widow, who did not speak English very well.

She had initially booked Sukhwinder and her two children on an Air India flight leaving from Toronto on 29 June, but her friend wanted to leave a week earlier and had visited the children's school, Moberley Elementary School, to ask permission for them to leave before the start of the summer holidays. It hadn't been easy to book a new set of seats for 22 June. It took several phone calls and Pradeep's considerable influence as a travel agent to get the three seats on Air India Flight 182, which was nearly full already. A day before her friend and the children were to leave, Pradeep also went to the Indian consulate on their behalf to pick up visas for the visit to India.

As she drove them to the airport through the heavy traffic on Marine Drive that Saturday morning, after the visit to the temple, Pradeep was still uneasy. Had Sukhwinder received a message of consolation for her past sorrows from the divine book? Could it be a reference to the miscarriage of twins suffered by the widow's sister-in-law the previous week after six months of pregnancy? Or was it a warning of future disaster? Still, Pradeep managed a chuckle as she glanced into her rear-view mirror at the children sitting in the back seat of her car. Her thoughts shifted to the visit of a friend to the Uppals' house the previous night. The friend, Sukhwinder's workmate, had brought gifts for the trip: a sari, a watch and a box of cookies. Nine-year-old Kuldip beat his sister to the watch, swiftly putting it on before Parminder could even move.

Later, after the visitor had gone, Parminder stood up and gave Kuldip a dirty look. She turned to her mother and demanded, 'Well, what did your friend bring for me?'

'She brought you this sari,' her mother said with a smile, showing her the garment.

'She should know I don't wear saris, I'm too young for that,' Parminder replied, looking woefully at the proud Kuldip, who was busy checking the time as if he had an important appointment.

They were eating breakfast by the time Sukhwinder's brother Major dropped in to say goodbye the next morning. As he entered, his eyes fell on the framed portraits of Parminder and Kuldip which sat on the television set. He apologized to his sister for not being able to see her off at the airport. He had to go to work on the early morning shift at the sawmill. Sukhwinder had then phoned Pradeep to ask her help once more, because the luggage wouldn't fit in the back of her sister-in-law's car.

At the airport, Pradeep helped her friend check in the baggage and then went over to another family which was also boarding the connecting flight, Air Canada Flight 136 to Toronto. She asked them to help her friend get around the busy airport at Toronto, explaining that Sukhwinder did not speak English very well. The children were jumping up and down with excitement over the summer holiday in India. But Sukhwinder was sad as she said goodbye to her 70-year-old father Mehnga Singh. She cried on Pradeep's shoulder as they hugged.

'Will you come and pick us up at the airport when we return?' Parminder asked Pradeep, who was mopping tears from her eyes. 'Should we phone you to let you know when we are coming back?'

'No darling, I already know when you are coming back,' Pradeep replied.

Then the three, mother and children, walked through the departure gate and disappeared. Pradeep would never see her best friend again, nor the children. There could be no escape now from their date with destiny. The gods had made their decision.

From another part of Vancouver, marine college instructor Sam Madon was on his way to Vancouver Airport to keep a promise he had made to his wife Perviz five weeks earlier. Madon had given his word that he would be in Bombay with her and their children Eddie and Natasha by 24 June, the date of his 42nd birthday. She had gone ahead in May for a planned vacation and a religious ceremony, *Navjote*, in Bombay for their son Eddie, who at nine was just old enough to be baptized into the Zoroastrian faith. There would be parties and a live band to herald the new era for Eddie, and he would now qualify to wear a special undergarment and a band around his waist to mark his status as a boy who was embracing his faith and becoming a man.

Madon had clung to his five-year-old daughter Natasha to comfort her as his wife and the children prepared to go into the departure lounge. The little girl had cried because Daddy wasn't going with her to India.

Madon's friend, Air India's Western Region manager Jehangir Parakh, also had been at the airport to see the trio off. He brushed his hand through Natasha's hair and said, 'Don't cry darling, Daddy will be with you soon.'

Madon hadn't made the trip with his wife and children in May because his vacation wasn't due to begin until the middle of June. After his wife and kids had gone, Madon insisted that Parakh should accompany him on his own trip.

'You'll get a free ticket anyway,' he said to Parakh. 'Why don't you come with me?'

Parakh toyed with the idea for some time. At one point he even called Toronto to tell the boss he might take a few days off to visit Bombay, and received clearance. But there was business to attend to in Vancouver and an audio-visual presentation planned on Vancouver Island to promote India to senior citizens, so Parakh told his friend he would have to go by himself. Sam boarded the Air Canada flight out of Vancouver with the Uppal family, never suspecting that less than 24 hours later Perviz would receive a phone call from family members in England to tell her that he wouldn't be able to keep his promise to be with her.

In a tiny village near the city of Ludhiana, in India's troubled Punjab State, a 70-year-old woman lay in bed, seriously ill. Doctors had told her family that her heart was weakening and she couldn't live much longer. The old woman kept asking to see her son Daljit Grewal, who had made his home in the Vancouver suburb of Surrey. She wanted to see him for one last time.

When the news arrived at his home in Surrey, 42-year-old Grewal, a sawmill worker, decided he had to go and see his ailing mother. As he prepared for his trip, though, Grewal had mixed feelings. On the one hand he was sad that he was going because his dying mother needed him. On the other hand, he was happy to be going home to the village where he grew up.

He took his wife Jagjit shopping to help him choose gifts, and bought clothes and a stereo cassette player for relatives back home. Electronic items are particularly popular in India, where the songs on All India Radio are sometimes the only entertainment to break the monotony of life and the struggle to earn bread.

'Look, you must study hard,' he said to his 12-year-old son Mundip before they left home to go to Vancouver Airport. 'I'll come back and see how well my son is doing at school.' He had similar words for his 10-year-old daughter Provjod as he hugged her at the Airport. Just then he had a thought. He walked over to the travel insurance counter and asked the clerk to give him air accident insurance for $400,000. His wife, looking over his shoulder, asked what he was doing.

'You might need it for the children's education,' he quipped with a smile.

He waved happily as he walked into the boarding lounge and passed through the security check for passengers, promising his wife he would bring back films of family members in the Punjab shot with his brand new movie camera.

As the Uppals, Madon and Grewal were checking in for their 9:00 a.m. flight to Toronto from Vancouver, another man was winning an argument with a Canadian Pacific Air clerk about getting his bag aboard Air India Flight 182. The man was insisting that his bag should be checked through all the way to India with Flight 182 although he himself only had a confirmed reservation as far as Toronto on CP Flight 60. There was a long, rapidly growing queue behind the man as he pressed the CP Air agent to do what he wanted. She finally gave in to his demand.

Air Canada Flight 136 and CP Flight 60 left Vancouver 10 minutes apart for the trip to Toronto. Both flights carried travellers who were on their way to connect with Air India and meet their loved ones in that far-off land, but the latter one, CP60, was also carrying the instrument of their destruction.

11

2

Prelude to Tragedy

Vancouver, 22 June, 1:15 p.m.

The veil of cloud that had hung over Vancouver International Airport all morning was parting slowly as the crew of Canadian Pacific Airlines Flight 003 received their final clearance for take-off for the long journey across the Pacific. The weatherman had been right again. He had promised that the clouds would part in the afternoon, giving the city renowned for its rain the same warm, sunny weather it had enjoyed for the past two days. As the sun burned away the last vestiges of the clouds, the cool, grey day turned into a bright one with pleasant 20° celcius temperatures. The massive Boeing 747 aircraft hurtled down the runway, its four engines thrusting with awesome power.

As the Jumbo soared to 30,000 feet over the sunny Pacific and turned west for the 10-hour journey to Tokyo, the 390 passengers and crew aboard relaxed. But one seat, allocated to an East Indian passenger in the Royal Canadian class, remained empty. The seat had been reserved for a man whose ticket identified him as L. Singh. The flight crew had waited for him patiently as announcements rang out in the terminal building asking Mr Singh to report to the boarding lounge. Finally the flight had left without him.

It was odd that Mr Singh hadn't made the flight. He'd paid $1,300 cash for his ticket. He'd even checked in his medium-sized, grey vinyl suitcase that morning. It was stowed in the belly of the mammoth aircraft. Yet when the flight left Gate

12

20, the clean-shaven, slender passenger who looked like a businessman was nowhere in sight. And so the Empress of Australia roared out over the Pacific, no one aboard suspecting that she carried a deadly cargo.

Toronto, 22 June, 4:30 p.m.

Things were running like clockwork at Lester B. Pearson Airport. The focal point for many travellers connecting with Air India 182, the airport saw a steady stream of planes touch down that afternoon. Air Canada Flight 136 from Vancouver, carrying the Uppals, Sam Madon, Daljit Grewal and a host of other passengers, had touched down precisely on time. And 20 minutes earlier, CP Flight 60 had arrived from Vancouver carrying a bag whose owner had stayed behind. The bag carried a tag which identified it as being destined for New Delhi. It was to connect with the flight to India.

A long walk faced the Vancouver passengers as they came out into the main terminal building and made their way towards the Air India counter. It was from here that they would check in for their plane, which was scheduled to take off at 6:35 p.m. – a two-hour wait.

As they headed towards the check-in, carrying their flight bags, another passenger had just made the link with the ill-fated flight. Rahul Aggarwal was having the time of his life in the airport lounge with his three university pals who had come to see him off.

A 23-year-old University of Manitoba arts student, Rahul had been his usual bubbly, unhurried self when he disembarked from a noon flight out of Winnipeg. He wandered out into the terminal, looking for his three life-long friends. But Raghu Rajan, Naiyer Usmani and Shaffique Dhamarjee were waiting at the wrong gate. After a few minutes of standing around, the trio realized their mistake and made their way to the spot where Rahul, looking impressive in a jacket and tie, was waiting with outstretched arms.

'Late again, eh!' joked one of his friends.

Rahul's eyes were beaming and his face showing the excitement of his trip to India. But he was serious when he replied: 'Yeah, man, I almost missed my flight in Winnipeg. Don't tell Dad I was late again!'

Over the weekend, his stepfather Arvind had taken Rahul aside to tell him he couldn't take his time where International flights were concerned. He had put his arm around the young man and given him a gentle reminder.

'I know you are a very busy man, Rahul, but an international flight won't wait for you – it's not like your personal train or something. So be on time, please.'

Rahul was anxious to please the older man. When Rahul was only 18 months old his natural father had died, but his stepfather had given him his all. The two were more like friends than father and son. During Rahul's visit to Thompson, Manitoba that weekend, the two men had talked for a long time. Rahul, always full of life and a young man who made friends as easily as he wore his smile, told his 'Dad' about his planned meetings with diplomats, scholars and government officials in India. It was the land he had left behind at the age of 14, but had never completely forgot as he shaped his new life in the country to which his family had emigrated.

Now Rahul was returning to India to write a thesis on Indo-Canadian relations as part of his final year's work for a Master of Arts degree at university in Winnipeg.

As the four friends sat in the airport lounge sipping their drinks and talking animatedly about their past together at the university, Rahul was in a philosophical mood. He was telling his friends that his real reason for going to India was to rediscover himself – to find out where he came from. To mingle with the crowds in Bombay's dusty streets and to breathe the air of a country that evoked pride in him.

Remembering what his father had told him about being on time for his flight, Rahul excused himself and went over to a telephone to call his home. But he didn't let out the fact that he had almost missed his flight in Winnipeg. Instead, he marvelled at the scene in the airport.

'I almost feel like I am in India, there are so many women in

14

saris at the airport today. I can't wait to get there.'

As the four friends reminisced, killing the hours Rahul had to
wait before boarding Air India Flight 182, the Boeing 747
that would take him on his pilgrimage was being readied by
ground technicians. 'Kanishka. Your palace in the sky' the
sign on the tail of the Indian jumbo said, and the plane was
indeed fashioned after fabled Indian palaces on the inside.
The jet was christened after Emperor Kanishka, who ruled a
state of India in the second century. The aircraft was being
prepared for a quick-turnaround flight back to India. It had
arrived at Toronto airport at 2:30 p.m. after only a short
refuelling stop in Frankfurt. Of the passengers it brought to
Canada, 68 were to continue on to Montreal with the same
flight, but all passengers had to disembark for immigration
and customs formalities.

On the tarmac at Gate 107, as ground personnel pumped
13,000 litres of fuel into the aircraft's huge tanks located
inside the wings, Air Canada maintenance staff cleaned the
cabin. Technicians were also busy attaching a fifth engine to
the port wing of the Jumbo. An Air India flight of 8 June had
returned to Toronto when one of its engines failed after
takeoff. The plane had borrowed an engine from Air Canada.
Now Kanishka was being fitted with the non-functioning
engine for the trip home.

There is nothing unusual about the carriage of an extra
engine, but strict rules have to be followed over the mounting
procedure for safety reasons. It wouldn't do to drop a three-
ton engine from 31,000 feet. For that reason, and because of
the time needed for the series of inspections that are made to
ensure safety, the process can be quite time-consuming.

Air India had informed Air Canada on 15 June that it
would like the engine prepared for shipment on the 22 June
flight, and foreman M. N. Patel had flown in from India to
supervise the loading. The extra engine was brought in from
Air Canada's hangar and installation began the moment the
flight arrived. The engine could not be carried on the wing
without first having its cowlings removed and the fans and

compressor rotors tied up. It would turn out to be a five-hour job.

Toronto, 5:15 p.m.

The mounting of an extra, non-functional engine to the left wing of the 747 wasn't the only cause of delay and frustration that evening as preparations were being made for departure of the flight. Inside the terminal's baggage transfer area, Burns International Security officers felt like kicking themselves too. A marvel of modern technology on which they relied heavily, the baggage X-ray machine, had packed up. There were hundreds of bags to be checked for the flight and the machine, with a mind of its own, was playing dead.

It wasn't the first time it had done that to Naseem Nanji and her co-worker Jim Post. The baggage scanner had done exactly the same thing on 8 June, the second time she had used it. Today, though, the machine had worked well enough when she first began using it around 2:30 p.m. At that time Air India had just opened its counters for passengers, and the flow of bags and cases was slow at the start. Naseem had put one cartful of baggage through with no problems, after moving the machine to its location and plugging it in as she had been taught recently. She had been given a crash course on how to operate it just 21 days earlier, after being moved from her normal beat near Canada Customs.

Now the machine had quit and the flow of bags was becoming much faster as the time approached for boarding the flight. Naseem looked at the huge pile of luggage and decided it was time to call in the experts. But Air India security officer John D'Souza couldn't fix the stubborn machine either, so D'Souza then asked Naseem and Jim Post to use a portable bomb-sniffer. D'Souza stood around watching Post use the PD-4 sniffer, then decided that a practical demonstration was needed. He lit a match and held it near the sniffer, to show Post how it gave off a long whistle when it detected fumes.

A short time after D'Souza left, a bag being checked by Post

16

made the sniffer beep. But the sound wasn't like the long whistle D'Souza had obtained with the lighted match. Post tried the device again on the maroon vinyl bag, and again it beeped when he passed it around the lock on the zipper. The beeping bag was allowed to go through, however, because it didn't produce the specified whistling sound. No one told D'Souza about the strange beep as ground crew began loading baggage into the forward and rear luggage holds of the aircraft.

Montreal, 5:30 p.m.

It wasn't so long ago that her son Mukul had dismantled the family's large and expensive cassette player while she took an afternoon nap, Subhashini Paliwal was thinking. Actually it had been almost eight years. But time had flown since the day the small boy had grabbed a screw driver and performed surgery on the cassette player, then appearing completely baffled as his child's mind tried to comprehend how to put it all together again. Now the boy was almost a man. Old enough to go to India on his own to see his grandmother in Agra, the city of the famous monument to love, the Taj Mahal. He was 15.

Mukul was sitting in the back of the station-wagon with his older brother Shailendra and younger sister Vandana as their father Yogesh manoeuvred the car into a parking spot at Mirabel Airport. The family had driven the 100 miles from Ottawa and had arrived with hours to spare, as Air India Flight 182 was not scheduled to depart until 8:35 p.m.

But Paliwal, a research scientist, believed in not having to rush at the last minute. The decision to send Mukul to India was his compensation to the ninth-grade schoolboy for having missed a trip in 1984 when everyone else in the family had gone to India.

Mukul was mature beyond his years. He loved the 'tabla' and often played the drums to accompany sitarists playing complex Indian classical music. Several times he had appeared on local television shows, to the delight of his

17

audience, thrilling them with the rhythms of an art that he had first learned when he was only nine. He also loved to make little electronic gadgets and repair calculators and cassette players. The gadgets he made, such as electric pinball machines, were greatly enjoyed by his friends at school. The fluently tri-lingual student – he spoke English, French and Hindi – was taking grade 10 courses in computers while only in grade 9. Only recently he had turned a pile of junk into a model train with the help of his father, doing all the electrical work including automatic switching on double tracks and lights.

At school Mukul was a better than average student, and had only missed the trip to India the previous year because he wanted to take further courses during the summer break. So he had stayed behind then and lived with his Uncle in Ottawa. In May, Mukul had been told by his father that he could go to India this time on his own if he wished. The teenager had agreed on one condition – that he could go again with the rest of his family for the planned marriages of two more of his uncles in December 1986. His father had agreed to the condition, and so the pair had booked a flight for 22 June.

When the ticket had been purchased, Mukul told all his friends about the upcoming trip. In the last week, the boy hadn't had much free time to spend with his family. First he was busy with exams in four subjects that he finished on Wednesday. Then for the next two days he was busy packing for the big journey. He included some presents for his grandmother, who was anxiously awaiting the arrival of the grandson she hadn't seen for more than two years. That morning, before the family left for Montreal, Mukul had insisted on paying a visit to his uncle's house.

At Mirabel Airport, as the family waited in the queue to check in Mukul's baggage, Yogesh Paliwal recognized another Ottawa family also lined up for the flight to India. Paliwal went over and shook hands with Satish Seth and his wife Sadhana. The Seths were travelling with their three children, including an infant who was still in a pushchair. Mukul's mother played with the child in the pushchair for a long time. Afterwards, Air India officials told the Paliwals that their plane would be delayed leaving Montreal. Paliwal asked anxiously why that was. 'We had to do some repairs in Toronto,' the official replied. The Paliwals headed out to a restaurant for some ice cream, but Mukul had never been fond of it and refused to eat. He preferred just to sit quietly through the hours of waiting and think about the short-wave radio he was leaving at home. There was a lot he would like to do with it when he got back.

Toronto, 5:30 p.m.

Rahul Agarwal and his friends had almost forgotten he had a plane to catch. Suddenly Rahul looked at his watch and decided it was time to go. But the flight was leaving from Terminal 2, and he was still in the other terminal. Rahul began running down the corridor towards the connecting tunnel. He didn't want to be late this time. Not for the flight to India!

But the young man needn't have worried. The flight wasn't leaving on time. And besides, he wasn't the only one who was late. The captain and crew of the Kanishka discovered that 59-year-old flight engineer Dara Dumasia had been left behind at the Royal York Hotel. The bus went back for the mild-mannered, quiet man when someone noticed halfway to the airport that he was missing. Dumasia, who had amassed 14,000 hours as an inflight engineer, had overslept, having missed the wake-up call for his last flight.

3

The Crew

Captain Hanse Singh Narendra eased himself into the commander's seat of the Air India 747 and began glancing at the maze of instruments. Flying was second nature to the 56-year-old son of a wealthy father from the Mathra district of the Indian State of Utar Pradesh. His other passion was hunting, a sport that had fascinated him since his childhood days during the British Raj. His parents couldn't keep him home as a boy because of his love for tracking down tigers at his uncle's huge estate in Mathra. During his school vacations and even now, when he had a few free days on his hands, Narendra would romp over to Sahanpur Estates where he had grown up. The huge parcel of land was still the focal point of the wealthy, close-knit family, even though the captain himself lived during his on-duty periods in a plush bungalow on a hill in what Bombay residents know as the Air India Colony in the Bandra section of this city of multitudes. Recently, he had also acquired land in Delhi, along with other Indian pilots, and was considering building a house. And his fascination with his uncle's estate had prompted the veteran pilot to buy his own ranch in Mathra. A few months earlier, during a trip to Ottawa, where his sister Sheila Mann lives, Narendra had told her of his dream of spending his retirement years at the ranch. His sister is the wife of Ottawa University professor Ranbir Singh Mann.

After finishing high school, Narendra had put his mind to

his passion for flying and had taken flying lessons at a private club. He became a commercial pilot in next to no time, and when Air India was formed from a number of smaller airlines, Narendra offered his services and the company hired him. This was in 1956, and those who had flown with him over the years knew him as a perfectionist who kept his cool even when the going got tough.

Captain Narendra knew the airplane he was sitting in today as well as he knew the back of his hand. But twice in the last year, Narendra had had his wrists slapped for putting too much faith in his co-pilots.

The first time he had slipped up was while flying over the territory of India's traditional enemy, Pakistan. It was 25 August 1984. Narendra was in charge of Air India Flight 1100 from London to Delhi when his plane, being controlled by a co-pilot, deviated from its track by 170 nautical miles. The commander, responsible for the safety of his aircraft and the actions of his co-pilot, was sent back to school for a few days, to re-learn instrument navigation systems and route cross-checking procedure. Like the able pilot that he was, Captain Narendra would not repeat that particular mistake.

The second error occurred when he was commander of a Delhi to Bombay flight on 6 December 1984. The runway in use at Bombay's Santa Cruz Airport was number 27, but the aircraft, again in control of a co-pilot, was seen approaching runway 32. Narendra's stern bosses sent him to a simulator to practise approaches and landings at runway 27 in Bombay. But that was all – there were no other blots on the captain's record. He was a check-pilot and had never recorded an accident in 20,000 hours in the cockpits of jets of all sizes.

Narendra was now waiting patiently to begin his task of flying the Kanishka to London and handing it over to another commander. He had flown in on Saturday 15 June, exactly a week earlier, from Frankfurt, Germany, hauling an extra engine that had been borrowed by Air India from Air Canada. This was not his usual route, but like his co-pilot Satwinder Singh Bhinder, he had been assigned to the flight because of staffing situations. He arrived in Toronto just one day before his girl-friend Valerie Margaret Evans flew in from London.

The tall, pretty, 43-year-old Air Canada passenger agent had met the witty, extrovert captain when he was flying to London almost 19 years earlier. Since then the two had developed a friendship that had grown into an affair.

It had been quite a week since Narendra had flown in the previous Saturday and gone to his suite at the Royal York Hotel in Toronto for a week-long layover. It was Valerie's birthday on Monday the 17th, and the captain had already reserved a suite for her next to his own when she checked in the next evening. Valerie and the captain had tried to meet whenever they could, having holidays in numerous cities and countries over the years, but their respective jobs had kept them away from each other more than they liked.

They spent the Sunday night together, mostly in their rooms at the Royal York. Then on Tuesday the pair left for New Jersey to spend a few days with their friend Saravjit Singh who lives in Spotswood with his wife and two children. It was something that they did without fail every year, and the couple didn't return to Toronto until Thursday. The next day the outdoorsy captain decided he would join the rest of his crew on a day-trip to breath-taking Niagara Falls.

Narendra was one year away from his retirement. And that's what he would talk about whenever he was with family and friends. He had brought up the subject in conversation with his sister in Ottawa when he spent five days at her home a few months earlier. At the time of that visit the trusted captain had been put in charge of flying a Jumbo to Canada for the inauguration of Air India's service to Toronto. On this latest trip to Canada, though, Narendra had not had a chance to visit his sister and brother-in-law, a chemical engineering professor, as they were travelling in the US.

At home in Bombay, where two full-size tiger skins adorned the walls of the living room, Narendra had left behind a 21-year-old son, Anil, who hadn't been well recently. Just before he left home, Narendra and his wife Sheila had discussed having him go overseas for medical treatment. But the couple had changed their minds, deciding that doctors in Bombay could do as good a job.

The only unpleasant episode during the week since

Narendra had arrived in Toronto had been the re-appearance of a 'pest' called Sharma. On Friday evening, the cool captain had had to use a stern voice in cautioning his younger co-pilot Satwinder Singh Bhinder, who was occupying a room in the same hotel, about Sharma. Valerie was listening when Narendra answered a phone call from Bhinder. The co-pilot said Sharma was in his room and wanted to see the captain. Narendra himself had received numerous messages from Sharma, whom he had met two years earlier while on a flight to Montreal. Since then, Sharma had continually left messages for him whenever he flew to Canada. On his inaugural flight in January, Narendra bowed to persistent pleas from him and visited his house, staying long enough to have coffee. But the captain had decided he didn't have much in common with Sharma, a resident of Toronto.

'Get rid of him,' Narendra said curtly to his co-pilot. Then Sharma came on the line, asking to see him, but the captain informed the man that he was busy and wouldn't have the time. Sharma sounded drunk on the phone, he told Valerie. A little irritated at the company Bhinder was keeping, Narendra passed the remark that his co-pilot, a fellow Sikh, was an 'extremist'.

But it was a remark made in the heat of the moment. Bhinder's loyalty to his country was never in doubt. On the contrary, he came from a family which, generation after generation, had served India through thick and thin. During its most difficult wars with arch-rivals Pakistan and China, Bhinder too had served his country – as an ace Air Force pilot – before he joined Air India as a commercial flyer.

That Friday evening, however, in the hotel room, Bhinder was having a tough time. His patience was being tested to the limit by the mysterious Sharma, but Bhinder still didn't lose his demeanour as a gracious host to the man who had called from the hotel lobby saying, 'You'll recognize me when you see me.' Somewhat baffled by the call, Bhinder had put his hand on the mouthpiece and asked his long-time friend Jagdev Singh Nijjar if it was all right if this man came upstairs. Nijjar, editor of a community newspaper in Ontario, shrugged. But neither of them were prepared for the man who staggered into the room smelling of booze and carrying a leaking bottle of whisky in his back pocket. The liquid was running down his trousers as he entered. Bhinder didn't recognize the man who now came in and greeted him like a long-lost brother. Then Sharma startled him by mentioning the model number of the aircraft he and Narendra were flying.

Bhinder looked at Nijjar. But his friend was no help, he just shrugged, slightly amused at Bhinder's plight. The guest flopped on a chair and Bhinder offered him a drink. Then Sharma stood up and asked Bhinder to come into the bathroom, where in a loud voice – Nijjar could hear what he said from the living room – Sharma asked the co-pilot to deliver an envelope to a Mr Khan in London, England. When they emerged, Bhinder suggested that they all go downstairs for dinner.

At the Black Night restaurant in the hotel's lobby, the hostess took one look at Sharma and decided he wasn't having any more drinks. Not at *her* restaurant, anyway. Bhinder had walked away to the gents' as Sharma and Nijjar sat down. Nijjar watched in bewilderment as Sharma began eating one bread roll after another. Nijjar was ready to blow his top, but kept the lid on his anger. What on earth was wrong with this man? Who the hell *was* he? Those were the thoughts in Nijjar's mind as Bhinder returned from the washroom and sat down. Sharma had by now cleaned out the basket full of rolls. Then, as soon as Bhinder sat down, Sharma stuffed a slice of bread into his mouth. That was enough, Bhinder decided. He wasn't going to put up with this ill-mannered idiot who had barged into his room. Not for a moment longer. Nijjar and Bhinder walked away to another table.

'Look at him now, he doesn't look drunk at all,' remarked Nijjar as Sharma suddenly walked out of the restaurant straight as a ramrod.

Nijjar and Bhinder didn't share the same political affiliations. Both were Sikhs by religion, but were diametrically opposed when it came to the future of India.

The Indian nation has been ripped apart in recent years by the turmoil in the Punjab State, which was home to both men

24

and to the rest of the world's 13.5 million Sikhs. The conflict in the Punjab, where the proud Sikhs ruled their own kingdom before the British Empire took over, has polarized Sikhs living in other countries like England, Canada and the US.

Some, like Nijjar, feel that a separate state should be carved out of the Punjab where the Sikh faith can flourish without the overwhelming influence of the majority Hindu population of India. But a huge majority still feel that this is not the answer, despite the bloody assault on the Sikh holy shrine called the Golden Temple by Indian troops trying to flush out extremists, and the incredible massacre of Sikhs in New Delhi following the assassination of Prime Minister Indira Gandhi.

Many Sikhs are still loyal to India, believing that the conflict can be resolved peacefully. Bhinder, like the rest of his family of staunch Indians, shared this moderate view. But that was politics and his friendship with Nijjar was not based on their political opinions. The two had met for the first time in Bombay while Nijjar was visiting the city and was introduced to Bhinder by a mutual friend. Hence on his first trip to Canada, Bhinder's first phone call was to the friend he had made four years ago.

On Thursday, the day before the incident with Sharma, Bhinder had paid a visit to Toronto's Punjabi market, a little Bombay which is a focal point of the city's large Indo-Canadian population and where everything from cooking oil to saris can be bought. You can bargain for things there, too, as you can in popular bazaars in Bombay. Then, in the evening, Bhinder called his cousin Harinder Kaur Mahal, wife of Vancouver area contractor Davinder Mahal. Bhinder was ecstatic over his first-ever trip to Canada, promising his cousin he would come back to the country as many times as he could. And he said he'd love to see Vancouver.

Harinder Kaur inquired about Bhinder's wife Amarjit and their two little children. During a conversation that lasted more than 45 minutes, Bhinder told his cousin that Amarjit and the two children had been with him in New York just three weeks prior to his trip to Toronto. But they had to go

25

back to Bombay, where the children were going to school. Previously, said Bhinder, he had been thinking about letting them come to Canada, where he'd pick them up. Harinder wanted to know all about their hometown of Karnal, in the state of Haryana, where Bhinder's father was a wealthy landlord and where Bhinder and she had grown up.

As he sat in the cockpit of Kanishka on 22 June 1985, Bhinder held a proud record of 7,489 hours of accident-free flying since joining Air India on 12 October 1977. Bhinder had arrived with the rest of Flight 182's crew on the bus delayed by flight engineer Dumasia's slumber. Valerie was at the airport too, and said goodbye to Narendra as he headed towards his aircraft to assume command. Some day soon, after he retired, Valerie was hoping the captain would marry her.

Toronto, 6:10 p.m.

The passengers for Flight 182 were pacing about in the boarding lounge when the announcement finally rang out that they could now proceed to their plane. There were 202 passengers, many of them children, boarding the aircraft from Toronto, including 21 who had arrived aboard various Air Canada flights from Winnipeg, Vancouver, Saskatoon and Edmonton during the day.

There was no sign, however, of the CP Air passenger from Vancouver who had made such a fuss with the clerk at Vancouver Airport to have his bag put directly aboard Air India 182 in Toronto. He hadn't even showed up to check in at the Air India counter there to see if he had a seat. But his bag was being loaded even as the passengers were boarding the airplane. Also, of the 68 transit passengers for Montreal who had arrived in Toronto with the flight from India and Frankfurt, only 65 had showed up for the final leg of their journey. The three missing passengers had probably used the well-known ploy of purchasing a ticket to Montreal simply because it was a cheaper destination from India than Toronto. It often happens that customers who are flying to Toronto will buy tickets for Montreal to save money, and then disembark in Toronto.

The security surrounding the boarding was extraordinarily tough. Passengers were first frisked under the watchful eyes of the Mounties and Burns Security while passing through the door-frame metal detector. Then they were again checked as they boarded the flight from the holding area. Security chief John D'Souza busily searched each and every piece of hand baggage being taken aboard the aircraft while metal detectors were used again to frisk the passengers. Furthermore, the airline, sensitive to the threat of terrorism, was also using a 'security numbers' system to make sure that all the passengers who had checked in were actually boarding the flight. Boarding of passengers was completed at 7:00 p.m.

The aircraft had already been delayed by 25 minutes. But there would be a further delay as ground technicians encountered a problem with loading of parts of the non-functioning engine inside the rear cargo compartment. To facilitate the insertion of the inlet cowlings of the fifth engine, Air Canada technicians removed the door fittings. Actual procedure as outlined in Boeing manuals calls for the removal of panels from the cowling to load it into the compartment. However, neither the Air Canada technicians nor Montreal maintenance manager Thiniri Rajendra were aware of the Boeing procedure. Rajendra, however, made sure the door fittings were properly reattached. He also carried out a final check of the aircraft and handed the certificate of air-worthiness to Captain Narendra, who duly accepted command of the aircraft.

Kanishka was now ready for take-off. The time was 8:16 p.m., a delay of one hour and 41 minutes from its scheduled take-off time. As it lifted off from Toronto's runway 24L with sandwiches and juice for the short haul to Montreal, the plane was carrying 270 passengers, 22 crew and hundreds of pieces of luggage, including two diplomatic bags from Vancouver. Also aboard for the hour-long flight to Montreal were security man John D'Souza, maintenance manager Rajendra and Divyang Yodh, passenger service supervisor, who had come in from New York because the regular agent in Toronto was on leave.

And, of course, the aircraft was still carrying the bag whose owner had never showed up. A bag no-one aboard Flight 182 was aware of.

4

Nightmare at Narita

Montreal, 9:10 p.m.

It had been a long, long wait for 15-year-old Mukul Paliwal.
The waiting was over now. The Air India plane that was to
take him to his grandmother and to the Taj Mahal had finally
touched down and parked at gate 80. Just ten minutes before
the flight arrived, Mukul had said goodbye to his family. In
the age-old tradition of showing profound respect for parents,
Mukul went down on his knees and kissed the feet of his
father and mother. Then he embraced them and kissed them.
His mother wiped tears from her eyes and gave him some last-
minute advice. He might be old enough to go to India on his
own, but he was still 'her baby', she said.

As Mukul and 104 other departing passengers proceeded
through the security check on their way to the boarding gate,
68 arriving passengers, including Air India staffers Rajendra,
Yodh and D'Souza, disembarked and were transported to the
arrival terminal.

There had been a few problems, however, while the 105
passengers were waiting to board. Burns Security officers
Real Gagnon and Jacelyn Cardinal had put aside a bag
belonging to a passenger boarding from Montreal. They had
spotted wires near the suitcase opening, and they weren't
about to take any chances, so they put aside the suspect bag.
They placed it next to the X-ray machine, which unlike the
one in Toronto was working fine. Then they picked out two
more bags which made them uneasy. The security men

informed Air India service agent Janul Abid of the suspect suitcases, but Abid asked them to leave the bags aside until D'Souza arrived from Toronto on the flight. While they waited, two diplomatic pouches were brought over from Ottawa by Mohinder Singh of the Indian High Commission. After the flight arrived, he and a ground worker proceeded to the aircraft and handed the smaller pouch, which weighed a kilogram, to flight purser Inder Thakur. The other bag, weighing nine kilos, was put with the diplomatic bags from Vancouver in a container in the forward luggage hold.

The aircraft, meanwhile, had been hooked up to a ground power cable to supply electricity because its auxiliary power unit, which supplies electricity when the engines are off, had been out of service since the aircraft left Bombay. Rajendra and three Air Canada technicians were busy performing a final check on the aircraft as CP Air catering staff loaded dinner and breakfast for the journey to London. Flight Engineer Dumasia had also come out of the plane to take a look at the fifth engine. He discovered that one of the latches was loose and asked an Air Canada technician to take care of the problem. It was soon rectified, and Dumasia, Rajendra and the three ground service men from Air Canada decided that there were no other snags.

D'Souza, meanwhile, had been informed by Abid that three suitcases had been laid aside as suspect baggage. The security man put the three bags through the X-ray machine and also used the explosives sniffer he had brought from Toronto. He decided to keep the bags in Montreal overnight. They could always be sent over on another flight if they proved to be harmless. But he and Abid made no effort to contact the passengers whose bags they had decided to keep.

Montreal, 9:50 p.m.

Mukul and fellow passengers boarding in Montreal were on their way to the waiting plane in a bus. Pretty sari-clad hostess Rima Bhasin and Sharon Lasrado, who had celebrated her 23rd birthday two days earlier in Toronto, were standing at the doorway to the Jumbo with their hands held together in front of them in the traditional Indian greeting *Namaskar*.

As the passengers from Montreal were seating themselves and putting away their hand baggage in the overhead compartments, a Mountie had been notified in the airport terminal that three suspect suitcases had been held back. He asked Air India representatives to come to the luggage holding area so the bags could be examined. But D'Souza was busy elsewhere.

Montreal, 10:18 p.m.

The time had come for Flight 182 to leave for its next destination. The microphone in the cockpit had just crackled and come alive with the voice of the air traffic controller. The co-pilot acknowledged.

Captain Bhinder: 'Air India 182. Good evening.'

Control tower: 'Air India 182, taxi to position, 06, 18.'

Bhinder: 'Taxi to position 06, 18.'

Control: 'Air India 182, airborne departure, 124.65 clear takeoff 06.'

Kanishka was off and away. Mukul was on his way to see his grandmother. But there was that Vancouver bag in the plane's belly that nobody knew about. It lay among 539 bags of all shapes and sizes, as well as the diplomatic bags and parts from the extra engine, that were stowed in the under-belly of the 747 as it took off with 104,000 kilograms of fuel, 307 passengers and 22 crew members for the flight to London.

As the aircraft disappeared into the night sky, back in the terminal building – more than 15 minutes after the flight had left – Abid and D'Souza walked into the baggage room where RCMP Sgt. J.N. Leblanc was waiting for them. The Mountie asked them to determine the owners of the bags, but the Air India pair told him the flight had already left.

The three bags were then taken to a decompression chamber owned by the airline with the toughest security in the world, El Al of Israel. There, dogs sniffed the bags and when they were opened later, the only unusual items found were an iron, a camera, a radio and hair dryer. Three passengers remained blissfully unaware that their luggage had aroused so much suspicion – and of the ironic fact that the cautious inspection of their harmless possessions could do nothing to protect them from the real danger that lay hidden in the hold beneath them.

The aircraft that was carrying the hopes and dreams of 329 people, Kanishka, a Boeing 747 model 237B, had been acquired by Air India on 19 June 1978 and had made its first commercial flight for the company on 7 July 1978. The plane was in its young middle age, with 23,634 hours of service completed in its seven years of operation. Its four engines were each capable of producing a roaring thrust of 48,650 pounds maximum. It was in topnotch condition, having been thoroughly serviced on 24 May, but it had had its share of problems which are considered common. The problems had been detected and dealt with promptly, however. Some examples from the plane's service history will show the nature of these faults and the repairs that were made.

13 July 1984, Dubai, Flight 868: The aircraft returned after aborting take-off due to an instrument showing no rise in engine pressure in the No. 1 engine. Checks showed slight wetness in bleed outlets, but no external oil leaks were noticed. Minor repairs were carried out and the plane took off without further problems.

18 July 1984, Delhi, Flight 105: The right-hand side fuselage skin in line with the forward cargo door was damaged by a high lift. Temporary repairs were carried out in Delhi and then permanent structural repairs were done at the Air India base in Bombay.

12 August 1984, Rome, Flight 135: The aircraft came in with its No. 2 engine shut down by the pilots because of dropping oil pressure and oil quantity. Checks revealed a leak from a cracked line. The line was welded in Rome and then replaced in Bombay.

24 October 1984, London, Flight 104: The aircraft suffered total loss of hydraulic fluid in one of the flap control mechanisms. Two of the four bolts holding the inlet pressure adaptor onto the flap control module had been sheared. Repairs were carried out and the flight continued to Bombay.

14 February 1985, Delhi, Flight 164: On arrival at Delhi Airport, a flap was found with damage measuring about 18 inches in length. The aircraft had apparently hit a foreign object while in flight. The flap was replaced in Bombay.

Tonight, as Narendra put the Jumbo into a steep climb and reached the cruising altitude, he could be certain that the plane was in tip-top shape. And it was carrying more than eight hours' worth of fuel, enough to take it to Paris in case an emergency forced it to overfly London. The passengers settled down for their dinner and inflight movie as Moncton air traffic control cleared the big jet to fly at an altitude of 33,000 feet. The plane's estimated time of arrival in London was 8:33 GMT, six hours and 15 minutes after take-off from Montreal. Over Gander, Newfoundland, traffic control on the ground gave the Boeing 747 oceanic clearance at normal cruising speed. But Bhinder advised them that he needed a slower speed because of the fifth engine. Gander cleared him to head out over the ocean at a reduced speed of Mach 0.81.

Narita Airport, Near Tokyo, 23 June, 6:20 GMT

Globe-trotting Woodwards Department Store merchandise buyer John Kennedy had three hours to kill before his trip home to Vancouver. He had just flown in from Taiwan after completing a shopping spree for the Vancouver store. After going through passport control at the gleaming Narita Airport, Kennedy had wandered into the huge waiting area where a maze of duty-free shops offered tempting electronic fare. But the 40-year-old buyer had had enough of shopping. Instead, he meandered over to a series of clear glass windows, stretching 60 feet in length, which offered waiting passengers a panoramic view of the hustle and bustle of one of the busiest airports in the world.

He watched CP Air Flight 003 coming in to the gate where he was to board it later for the flight home. When you travel a lot as Kennedy does, three hours to spare at an airport like Narita isn't an awfully long time to wait, and he was used to it. Now, as he looked out through the sparkling glass windows, he could see little forklift trucks driving up to the Canadian Pacific jet and driving away again with containers of luggage for the passengers who had just arrived from Vancouver. With Pacific head-winds not as harsh as usual, the orange and white CP Boeing 747 had made good time, coming in ten minutes ahead of schedule at 2:45 p.m. Tokyo time.

Looking down through the windows, Kennedy could see the busy little baggage trucks disappearing beneath him into the ground-floor terminal where handlers would sort out the bags and cases before passengers picked them up. Deciding at this point that he had better things to do than just watch the baggage trucks or stare into the distance at the sleepy little city of Narita, Kennedy turned away from the window.

He hadn't taken two steps when a thundering blast shook the whole airport. The force of the explosion almost knocked him off his feet. What the hell was that? he thought, as his heart pounded. It sounded like a cement truck had come crashing through the roof of the terminal. The time was 3:20 p.m. There was a moment of stunned silence in the airport as the roar of the blast died down. Then came a second explosion, this one of human activity, of babies crying, women screaming and men running.

All hell seemed to break loose as Kennedy stood rooted to the spot. He waited, expecting to hear another massive blast. But none came. From the ventilation grille behind him, thick acrid smoke began filtering into the lounge. This leisurely transit stop was turning into a nightmare, Kennedy thought. This was a bomb, he was certain. It had to be; nothing else could make a sound like the earth falling apart.

Amid the chaos and screams, sirens wailed from the tarmac below as Japanese police hurled themselves into action. From the windows Kennedy could see ambulances, firetrucks and police cars dashing towards the scene of the explosion on the ground level, right below where he was standing. The dismayed Canadian watched the scene on the tarmac below in awe, his mouth hanging open. The first two victims were being brought out on stretchers. Mercifully, their bodies were covered from head to toe. Obviously they were dead. Then two more were brought out, their bodies covered only in blood. The ambulances sped away, sirens wailing, and more arrived to carry two more casualties. These baggage handlers too were relatively fortunate. Blankets did not cover their heads. They were bloodied but still alive.

As the minutes passed the chaos began to subside. From where he was standing, Kennedy had a panoramic view of the incredibly efficient Japanese security men in action. An armoured car drove in to probe the rest of the baggage that had been taken off CP 003. The bomb squad was taking no chances with the plane that had brought deadly cargo from Vancouver. Kennedy then saw the grimly humorous figure of the chief of security, shouting orders to his men; behind him an attendant was carrying a pole with a large number one to identify the man in charge. In chaotic situations, it is imperative to know just who is issuing the orders. Also on the tarmac, Japanese press officers were briefing reporters, drawing out chalk-board diagrams for the curious newsmen. Was it a bomb? Where did it come from? How many dead? How many wounded? Had anybody taken responsibility yet? The officers had few answers right then. One thing they were sure about was that it was a bomb. And it had been brought to Narita by that orange and white Canadian Pacific jet.

Already the Empress of Australia, as the plane was called, had become infamous. She sat on the tarmac as tourists took pictures by which to remember the terrible drama that had been played out in front of their eyes. It was all like a bad dream. Particularly bad for John Kennedy, knowing that the vehicle of death, the aircraft sitting on the tarmac, was to become Flight 004 and carry him home across the Pacific.

Down on the ground level, firemen had doused a small fire and forensic experts were busy clearing up and photographing the mess. The deadly blast had occurred just as luggage handlers were removing a bag from a transit container. It blew the men several yards away. Particles of metal and plastic, thousands of them, were scattered around as if a hurricane had hit the terminal. The bomb had made a crater in the concrete floor. Police had blocked all roads leading in and out of the airport and vehicles were being checked before leaving or entering the terminal area. A task force of three dozen policemen was formed immediately to probe the explosion.

Some distance away, Air India Flight 301 to Bangkok was being refuelled when the explosion ripped through the terminal. No one knew it right then, but the bag that had blown up was to be loaded on that plane, whose destination was the supposed final destination of the missing passenger L. Singh. The aircraft – and probably many lives – had been spared.

The trip home for Kennedy was delayed while security officials went through the CP Empress with a fine-tooth comb. And as Kennedy boarded the plane, still a little apprehensive, he and the other passengers were physically searched.

It was a grim flight. Most of the way, he pondered the macabre scene at the airport and the awesome power of the explosive. What if the bomb had gone off aboard the CP plane while it was in flight? No aircraft could survive a blast like that. What if the plane had been late, just 30 minutes late? He didn't know it then, and the captain wasn't going to tell his nervous passengers, but another tragedy, one of massive proportions, had happened while the CP Empress was still on the ground in Tokyo.

'There are some unfortunate things happening,' allowed the captain over the cabin microphone. 'But everything will be okay.' Kennedy didn't know it, but the captain had just made a heavily veiled reference to a nightmare that had happened thousands of miles away.

5

Countdown to disaster

The Atlantic, 23 June, 6:20 GMT

The horrifying Narita Airport bag bomb carried by CP 003 from Vancouver had just exploded. But thousands of miles away, at an altitude of 31,000 feet over the North Atlantic, there was no sign of trouble aboard Kanishka. Four hours had passed since the Boeing 747 lifted off from runway six at Montreal for the six hour and 30 minute flight to London.

Passengers were now watching a Hindi movie with a special bonus – a flight purser who was also a movie star present among them. It was Inder Thakur, who at 35 had already earned distinction not only as a screen star but as an international model and fashion designer. The passengers were indeed being treated according to Air India's promise: to make them feel as if they had already arrived in India from the moment they boarded. With pretty sari clad hostesses serving Indian food in an aircraft decorated like a palace from princely days of the British Raj and before, and an Indian actor too, they could hardly feel otherwise.

Thakur, a 13-year veteran of Air India, was today flying with his wife and child. Only the month before, the versatile purser had shown off his exclusive Indian designs at a convention of the World Modelling Association in New York. The smiling charmer had walked away with the award for the International Fashion Designer of the Year. *The Magic Carpet*, Air India's magazine, carried a picture of Thakur with WMA president Ruth Tolman.

There had been four hours of uneventful flying, as they would say in pilot talk, and not an iota of turbulence. From the cockpit of the 747, the pilot could see the sun rising over the Atlantic, and way below, at 10,000 feet, cloud blanketing the sea. The weather was no problem, either on the surface of the Atlantic or at flight level of 31,000 feet where Kanishka was cutting through the air at speeds varying from 287 to 296 knots – slightly slower than normal because of limitations imposed by carrying an extra non-functioning engine. The surface temperature was 13 degrees celcius with a wind speed of about 15 knots. There was no cumulonimbus cloud or thunderstorm activity. At the height Kanishka was flying, the temperature was a chilly -47 degrees celcius with a steady westerly wind flow. Everything about the aircraft was ship-shape, too; there was no sign of anything wrong mechanically.

There was nothing that Captain Narendra could complain about. Nothing to prevent the Jumbo from touching down at London's Heathrow Airport at 8.33 GMT as scheduled. But the captain, crew and passengers of Air India Fight 182 didn't know yet that the countdown to disaster had begun.

7:00 GMT

There was just an hour and thirty minutes to go now before the aircraft was to touch down at Heathrow. Just time enough for breakfast and the collecting of the trays before passengers readied themselves for a stop-over in London. The plane was to begin final approach in about one hour and 15 minutes. Most of the passengers sat without their seat belts fastened and in the cockpit, flight purser Jamshed Dinshaw, who had continued working for the airline despite a 1978 Air India crash in the Arabian Sea that took the life of his sister, was standing with inflight supervisor Sampath Lazar. Dinshaw, travelling with his wife Pamela, a hostess on the flight, was listening to small talk between Bhinder and Flight Engineer Dara Dumasia. The cockpit Voice Recorder was churning away in the back of the plane, keeping an ear on the conversation of the cockpit crew.

'Dinshaw,' called Bhinder from the co-pilot's seat.

'Yes, sir,' answered the bearded purser.

'Do me a small favour,' said Bhinder in English.

'What's that?'

'*Edkam* end *pe*, 54 seat *pe*,' (at the very end of the plane, on seat 54) said Bhinder, mixing his Hindi and English. 'A boy is sitting there. Inder Thakur knows. He just wanted to have a look in the cockpit.'

'Where is he?'

'Inder Thakur knows about him,' Bhinder repeated.

'Okay, 54 seat. Can I send him now?'

'After about 15 to 20 minutes,' replied the co-pilot, not knowing that he wouldn't have that much time to do a small favour for a little boy who wanted to look at the dazzling array of instruments in the cockpit of a 747.

There was silence in the cockpit for about five minutes, then Bhinder spoke again, seemingly unhappy about all the beer the girls had brought in to take home from their week-long layover in Toronto. There is nothing so popular as gifts of foreign beer in India.

Lazar replied, calling it a hard core problem. Captain Narendra, known as an extrovert with a remarkable wit, had remained silent so far. But now he joined in the conversation too, saying that some of the cabin crew might be carrying beer for others as well.

'Somebody ten beers, somebody six beers,' said Bhinder. 'Hold, hold.'

It was time to check with Shanwick Control, a joint air traffic control centre in Ireland between Prestwick and Shannon. The time was precisely 7:06:39 when Bhinder made contact.

Bhinder: 'Shanwick AI-182. Good morning.'

Shannon: 'Station calling Shannon go ahead again.'

Bhinder: 'AI-182. AI-182 is 51 N 15 W at 0705 level 310. Estimate FIR 08 W 51 N 08 W at 07:35.'

Shannon: '182, your correct Shannon frequency is 131 15.'

Bhinder: '131 15, sir.'

The cockpit Voice Recorder next picked up Shannon communicating with TWA Flight 770, which was following behind Air India 182 by a distance of five miles and flying at an altitude of 35,000 feet. Trailing the TWA plane by about 20 miles was CP Air Flight Empress 282. Shannon cleared it to fly to Amsterdam at an altitude of 37,000 feet.

At 7.08 Bhinder contacted Shannon again to report his position. Shannon asked him to 'squawk 2005.'*

'Right, sir, squawking 2005, 182,' Bhinder duly replied.

With the communication with Shannon over, engineer Dara Dumasia called Bhinder. Dumasia said the flight purser wanted some 30 seals for the bar, for customs inspection purposes in London, and asked if he could call London Operations and tell them to have the seals ready for the flight's arrival.

Bhinder: 'Customs.'

Dumasia: 'Custom seals. *Wo bar seal karane ke liye* (seals to close the bar). For their arrival – customs. Bar ...'

That was the last time anyone talked in the cockpit of Kanishka. The sound of an explosion interrupted the conversation. The time was 7:14:01. It was the end.

The massive Jumbo went into wild gyrations, breaking up in the air and hurling bodies, luggage and metal everywhere. It was all over for Uppal and her two children, for Madon and his promise to his wife, for Rahul Aggarwal, for Daljit Singh Grewal and that boy wonder Mukul Paliwal from Ottawa who was on his way to see his grandmother. And for Captain Narendra and co-pilot Bhinder. Most of the 329 passengers and crew were dead before they hit the water, having been wrenched violently out of the aircraft's sheltered cabin and exposed to the viciously cold and oxygen-starved upper air. Their bodies were twisting around and their limbs flailing wildly as they hit the Atlantic. But some were still alive, although unconscious, when they hit the water. They drowned. It was an accident where no one stood a chance to survive.

It had been four hours and 56 minutes since the aircraft had left Montreal. The first item that had been liberated from the

* *Footnote:* 'Squawk 2005' – a code supplied to the aircraft so that the craft's transponder corresponds to signals from ground radar.

bottom of the ripped aft luggage compartment was a bag and portions of the fuselage skin. The bag slowly sank to the bottom of the ocean, with clothes protruding from its torn vinyl skin.

Search and Rescue, 7:14 GMT

As the remains of Kanishka sank to the bottom of the ocean, Shannon Air Traffic Control was still in the dark over the fate of the aircraft with which it had communicated just four minutes and ten seconds ago. There was no 'Mayday' call. No emergency was declared. But as Air India Flight 182 was falling out of the sky, a few milliseconds of microphone clicking sound was picked up by two alert traffic controllers in Shannon, M. Quinn and T. Lane. Just after the burst of microphone sounds, Air India 182 disappeared from Shannon radar.

Lane looked again at his radar, expecting that the missing blip would return, but nothing happened. The flickering diamond shape just wouldn't come back on the screen. The frantic controllers called Air India 182 five times at 7:16 but there was nothing except silence.

At 7.17 Shannon contacted TWA 770, which had been flying just behind Air India but at a higher altitude.

'Okay, just calling to tell you there's an Air India there with you and . . . he's not talking to us at this time – would you just give him a call please?' Quinn requested.

TWA 770: 'Air India one eight two from TWA seven seven zero.' There was no reply.

'Ah, Shannon, TWA seven seventy. He won't answer us either.'

'We just had him ahead of you there,' said Shannon. 'Five miles just ahead of you there and his squawk is gone off the scope . . . that's why . . . Can you see anything ahead of you there? . . .'

TWA 770: 'Well, no, we don't see him.'

Shannon: 'Air India one eight two, Air India one eight two.'

41

TWA 770: 'We've been looking and I've been calling him on guard' – an emergency frequency monitored by all aircraft – 'and he hasn't answered and we don't see him.'

Shannon then asked CP 282 to see if the missing aircraft was visible anywhere. But CP 282 could only see the TWA jet. Next Shannon asked TWA to descend and turn to see if it could spot Air India 182. The control centre then asked other airlines in the area to keep an eye out. But it was as if the lost plane had never existed.

7:30 GMT Just 16 minutes after AI-182 disappeared from radar and lost radio contact, Shannon informed the Irish Marine Rescue Co-ordination Centre that the Indian Jumbo en route from Montreal to London had disappeared. Shannon asked for emergency action. Ten minutes later the marine rescue service called the Valentia Coast Radio Station to alert all marine vessels to keep an eye out for any sign of wreckage, giving the last known location of Air India 182 as approximately 180 miles south-west of Cork.

7:50 GMT The Irish Naval vessel *Aisling*, a warship, reported to Valentia Radio that it was 54 miles from the last known location of the flight and was proceeding in that direction. The Irish Naval Service and the Irish Army Air Corps were then briefed about the emergency. A distress message was broadcast again for all shipping in the area.

8.04 GMT The Panama-registered vessel *Laurentian Forest*, owned by Federal Commerce of Montreal, reported it was 22 miles away from the distress area and would head towards it, The cargo vessel's captain inquired if other ships were in the area, and was told about the *Aisling*. Marine rescue Co-ordination Centre in Shannon then alerted the rescue co-ordination centre in Plymouth, and they advised Shannon that a Nimrod rescue aircraft was being readied for departure shortly and Sea King helicopters were already on their way to the quiet sea-side community of Cork for deployment.

8:29 GMT Shannon advised Valentia Radio that aircraft passing over the site were picking up an Emergency Locator Transmitter signal at location 51 N 15 W and all ships in the area should be requested to report to Valentia. Three other ships in the area, *Ali Baba*, *Kongstein* and *Western Arctic* then reported to Valentia and offered to move to the location of the stricken airliner.

9:05 GMT The *Laurentian Forest* advised that it was about five miles from the site but couldn't see anything. Three other ships were also steaming to the site of the SOS and four Sea King choppers were now en route.

9:13 GMT Eight minutes after its call to Valentia to say that it was unable to spot anything, the *Laurentian Forest* was on the air again. It had a grim report. It could see what appeared to be life rafts about two miles from its position. As the ship headed towards the area where the life-rafts were, it spotted more wreckage from the downed airliner, and reported that the rafts were not inflated.

9:37 GMT As the cargo ship passed the uninflated life rafts and debris, it made another find. Three bodies could be seen floating in the cold waters of the Atlantic. The time in Vancouver was 2:37 in the morning, while in Montreal it was 5:37. Most Canadians, including the victims' families, were still sleeping, unaware that loved ones they had said goodbye to only a few hours earlier had died. The three bodies located by the *Laurentian Forest* were picked up. They were naked, with limbs twisted around like a rag doll's.

9:45 GMT By now, one of the largest search and rescue operations in history was underway. MRCC Shannon had decided that Cork, for operational and security reasons, would become the operational base for the search mission and Air Traffic Control at Cork airport was informed of the decision. An hour later, a prohibited zone for aircraft was established within a radius of 40 miles from the scene of the disaster. Valentia Radio then assigned specific quarters for each ship to search, with all bodies and wreckage found to be taken aboard and a report filed with the *Aisling*, which was established as the command ship for the huge Atlantic search.

11.33 GMT As the *Laurentian Forest* began fishing bodies out of the water, eight Spanish trawlers were headed for the search site. The deck of the *Forest* had by now begun to resemble a morgue, with 66 bodies, some naked, laid side by side. The ship *Star Orion* had by now offered to refuel any vessel needing it in the area of the search. The *Aisling* had also picked up more bodies by then. The count would go up to a total of 130 bodies aboard various ships by nightfall as the search wound down and the 'Mayday' alert was downgraded. **18:40 GMT** By this time 64 bodies had already arrived at Cork airport and five pathologists were called out to Cork Regional Hospital for the autopsies. One of the items amongst the wreckage brought to Cork was a child's Cabbage Patch doll, found floating amidst the scene of disaster and death.

Cork would not be the same for months to come. Investigators from India, the United States, Britain and Canada, as well as grim faced relatives hoping to provide a decent burial to their kin, would begin to swamp the town the next day as the search for more victims and debris continued with a steady stream of planes and helicopters landing and taking off from the airport.

When the search was finally over and the grim statistics had been totalled up, the Air India tragedy had become the fourth worst aviation disaster in history and the worst ever for Canada, as 280 of the 329 victims were Canadians. Later, history would also record it as the bloodiest terrorist act of its kind in modern times – an attack that was launched from Vancouver.

Most people in Vancouver and Montreal were still fast asleep when the first bodies of the victims of the Air India disaster were found. Radio stations were already broadcasting the news that the flight had disappeared from radar screens at Shannon and had possibly crashed in the Atlantic.

It was 5:30 a.m. when the telephone rang at the Ottawa home of Yogesh Paliwal. Was it Mukul calling from London to tell him he had arrived on the first leg of his journey? Who else, Paliwal asked himself, would call at 5.30 in the morning?

It wasn't Mukul. The caller was a friend who had been listening to the radio. The flight on which Mukul was travelling had disappeared, Paliwal's friend was saying. Paliwal got out of bed and woke up his wife. She began to wail at the news. Paliwal, his hands shaking, grabbed Mukul's favourite short-wave radio and tried to tune in the British

Broadcasting Corporation from London. His son's radio brought him the news. The son who had kissed his feet in the traditional Indian way of showing respect for elders less than eight hours earlier was dead. The radio said there appeared to be no survivors.

Just about the time Paliwal was tuning in the radio in Montreal, the telephone was ringing at the Vancouver home of Air India's Western Region manager, Jehangir Parakh. Parakh had been out till late that evening partying with friends. He was fast asleep when the telephone rang at around 2:30 a.m. Vancouver time. His wife picked up the phone. On the other end of the line was Air India sales representative Derek Menzes from Montreal.

'Just take a message,' said Parakh to his wife. 'I'll call him back tomorrow.' She said he insisted that the matter was urgent.

Still half-asleep, Parakh picked up the telephone and heard Menzes saying, 'Flight 182 has disappeared from radar. Your phone will be ringing off the hook tomorrow.'

'God bless you,' said Parakh and hung up, thinking that there was a routine technical problem of some sort. He is still kicking himself and wondering why what Menzes was saying didn't register on him right then. But he was completely exhausted after having worked all day and then gone to the party in the evening.

At 6:30 in the morning, Parakh's son shook him awake.

'Dad, something has happened to the Air India flight – it's hit a radar station or something.'

Parakh picked up the news of what really happened to Air India flight 182 on morning television news and headed straight for his office at Air India's downtown Vancouver headquarters on West Broadway. A telex was waiting for him there which read:

'On behalf of Air India, I regret to have to advise you that one of our aircraft, VT-EFO Flight 182 of 22 June from Toronto and Montreal to Delhi and Bombay via London, was reported lost at sea off the coast of Ireland in the early hours of the morning.

'The 747 aircraft lost contact at approximately 0715 GMT, 180 miles from Cork, Ireland.

'The total number of passengers on board was 307, plus 22 crew. The latest information available is that the wreckage has been sighted and several bodies have been picked up. There are no reports of survivors as yet. The passenger list will be released once next of kin have been informed. We have further been advised that some more bodies have been sighted.

'The last contact with the aircraft was 0710 GMT when the aircraft was cruising at 31,000 feet. Conditions at that time were reported as normal. The control tower lost contact with the aircraft at approximately 0715 GMT.

'The commander of the aircraft was Capt. Narendra. He joined Air India in the year 1951 and has been a senior commander since 1964.'

'God, 329 people!' said Parakh to himself. 'Sam Madon was on that flight – *I* could have been on that flight.'

Crying relatives and anxious friends began pouring into the Air India offices in Vancouver, Toronto and cities across the world. In Bombay a crowd gathered outside the Air India office near the scenic Marine Drive. Sheila Narendra, wife of the commander of the aircraft, was among those people. She knew then that her husband had died. Earlier, Air India representatives had come to her house, saying only that the flight had gone missing. Special telephone lines were set up by Air India in several major cities to relay news of the tragedy to families. A telephone call from London to Bombay told Perviz Madon that Sam wouldn't be with her for his birthday or for Eddie's *Navjote*. As Sam's younger brother Cyrus left Vancouver for London, Perviz was making her way to England, too, from the other side of the world. Three days later, Major Sidhu was on his way to identify the battered body of his sister Sukhwinder and little Parminder. Their bodies had been fished out of the Atlantic with 129 others found floating among the debris of Kanishka.

As Sidhu boarded his flight that day, security at Vancouver airport had been beefed up dramatically. There were X-Ray machines to check baggage where there had been none previously, and a dozen policemen, some in plain clothes, mingled with the crowd. It was a grim flight for Sidhu. He hardly said a word during the flight. Most of the time, he sat looking out of the window.

Also on the flight was Cyrus Madon. He talked with fellow passengers on the flight.

'You know,' he said, 'I always thought of my brother as a happy man, but now all I can think about is him falling from the aircraft. All I am getting is a still picture of him falling from the sky.'

In Cork, Ireland, Sidhu was shown pictures of Sukhwinder and Parminder. They didn't look anything like they did when he had seen them on Saturday morning to say goodbye. Sidhu will never forget the missing eye, the broken nose and a huge stitched-up cut on Parminder's face.

'I couldn't be positive it was them. I was only 50 per cent sure – then we had to use dental records for positive identification.'

The body of Kuldip, the favourite grandchild of Mehnga Singh Sidhu, was not among the 131 bodies picked up within two days of the disaster. The 70-year-old man sat in one corner of his home in Vancouver, gazing at the ceiling with a blank look in his eyes. Destiny had dealt him more blows than he could bear.

As Canadian families counted their massive losses, Prime Minister Brian Mulroney was busy sending his condolences to Indian Prime Minister Rajiv Gandhi, to the chagrin of the victims' relatives in Canada. Two hundred and eighty of the victims were Canadians; it didn't matter that most of them were of Indian descent. The Prime Minister quickly realized his mistake and hastened to correct it.

'I would like to convey my deepest personal condolences and those of the Canadian government to all members of the bereaved families on the occasion of the terrible loss of Air India Flight 182,' the Prime Minister's message to the Canadian relatives said.

*

In the Air India office during the morning of Sunday 23 June, as word of the scale of the disaster filtered out, was acting Vancouver Indian Consul Gurinder Singh. Both he and Parakh had been at the party together the previous evening. An early morning call from his superiors in Ottawa had asked the Consul to begin looking into the tragedy of Air India, as there was some suspicion of sabotage because of the Narita incident. Amid the gathering of relatives and the constant buzz of telephone calls, the telex machine in the Air India office began whirring again. It was Air India's Tokyo office.

'Good morning,' said the text. 'Kindly investigate details of following two passengers supposed to travel CP 003 22 June Vancouver/Narita and Air India 301 23 June Narita/Bangkok, who no showed Air India 301. Singh/A Mr PNR H6269 original CP/Vancouver booking reference YVRCP/QG4JIBA301XF of 16 June and Singh/L Mr PNR H639R original CP/Vancouver booking reference YVRCP/UZCJPSA301XF of 20 June 85. Most grateful your urgent reply due above required by manager Japan in connection with Narita Airport baggage explosion. Regards.'

The mystery of the missing Singhs had begun.

PART TWO

THE INVESTIGATION

1

The Investigation Begins

It was supposed to be John Kovalick's day off. The 47-year-old staff sergeant from the Richmond detachment, Royal Canadian Mounted Police, was enjoying the peaceful Sunday morning in his home not far from Vancouver International Airport. The quiet of the morning was disturbed only by the occasional roar of a big jet taking off or landing. But there wasn't to be a day off for Kovalick, a soft-spoken veteran policeman – not today. He'd heard the news of the bomb explosion at Narita Airport and the larger tragedy of Air India flight 182 on the morning radio news. Still, when the call came in from his office, it took him by surprise. The head of the Richmond detective section had just been handed the biggest assignment of his career.

Kovalick wasted no time in getting to the station. As he drove the short distance to his office, through the nearly deserted streets of this slumbering suburb – generally regarded as a 'bedroom' of neighbouring Vancouver – he thought about the task that had been landed in his lap. A criminal investigation into two aviation tragedies, half a world apart, in of all places, Richmond. It was unheard of. The veteran cop had probed many criminal cases in Richmond. Stabbings, murders, robberies, rapes. But nothing like this. Ever.

Driving into his reserved parking spot at the Mountie office on Minoru Boulevard, Kovalick went upstairs to be greeted by Inspector Bruce Giesbrecht, who had also come in from a day off. Together, Kovalick and Giesbrecht, chief of operations, sat through a briefing in which the sketchy details that were available were laid out.

From Japanese police, the Mounties had already learned that the explosion had occurred in a bag off-loaded from CP Air Flight 003, which had made a non-stop flight from Vancouver to Tokyo. The Japanese police had quickly seized the passenger list and held those who had arrived aboard the flight. But it became clear to them that nobody in his right mind would board a flight knowing that a bag bomb was aboard. It had also become clear quite quickly from Air India officials that two transit passengers, A. Singh and L. Singh, who were supposed to board Air India Flight 301 to Bangkok two hours after the CP Flight landed in Tokyo, had not shown up. It was more likely that the bag belonged to one of the no-show passengers.

At the briefing in Richmond, the mysterious destruction of Air India Flight 182 was so far the bigger question mark. But theories of sabotage and of a possible link between the Narita explosion and the Air India tragedy were already the subject of speculation. For the moment, though, Narita had the immediate proof of sabotage – most likely launched from Vancouver. They had physical proof in the form of two shattered bodies of Japanese baggage handlers and thousands of pieces of debris. They also had the lead on the two missing passengers.

There were other reasons, however, for launching an immediate Canadian probe into the disasters. The credibility of the Canadian transportation system as well as its security were at stake. The public had to be assured that it was safe to travel to and from the country. And because the airport fell into the jurisdiction of the Richmond detachment, the probe would have to begin with Kovalick and Giesbrecht, a 28-year veteran of the force. As the briefing wound up, both men shared the same thought. Nothing like this had ever happened in Canada before. It was a type of criminal investigation that had never been undertaken in Canada – one, possibly two, terrorist acts which had taken a total of 331

51

lives.

As the detectives from the Richmond general investigation section began their work that afternoon, Kovalick and Giesbrecht could be sure of only one thing. They were in the middle of making history.

Tangled Tickets

From the beginning of the investigation one day after the disasters that rocked the world, it was obvious that CP Air reservations and sales agents would have a story to tell. At least they could tell the Mounties something about the two Singhs missing at Narita Airport. The first job, after the briefing ended, was to track down the CP Air personnel who had vital information. This wouldn't prove too difficult, for most of the agents involved had already put two and two together and were busy discussing among themselves curious incidents that had happened to them hours before the tragedies.

The most vital early interview would be with reservations agent Martine Donahue, who had booked one of the Singhs who failed to show up at Narita. But the RCMP had something else going for them, too. Every airline ticket has its own biography. The time it was booked, the time it was picked up and paid for, the telephone numbers of patrons or their place of contact and even previous route selections and cancellations are kept on a computerized file called a PNR. Even more information is available from computer data banks used by most major airlines nowadays.

It was while going through passenger manifests that they learned that a third passenger using the name Singh had also gone missing. This one was supposed to have taken CP Flight 60 out of Vancouver on Saturday and to have connected with Air India Flight 182 in Toronto. As far as the detectives could tell in the early hours of the probe, one of the two Singhs missing in Narita, A. Singh, had made a reservation for the flight to Tokyo but had never purchased a ticket. However, the two others had. Both were eastern Indian males. Both had given the name Singh as their surname. It had to be more than sheer coincidence. This would be the first real clue for the Mounties in piecing together the story of what had happened to Air India Flight 182. But who were the Singhs and how had they made their bookings?

The answer to this question lay at the Bental Centre. A tall office tower in the heart of Vancouver's central business district, the fanciful glass and steel tower is home to CP Air's booking operations. Within hours of their Sunday-morning briefing, the Mounties were being shown a bewildering maze of reservations and cancellations for Singhs flying on both Air India and the CP Air flight to Narita which had carried a bomb across the Pacific.

The file would be a revelation to the Mounties. The maze of bookings and cancellations appeared to indicate that those who had booked the flights weren't certain of where they were flying to or when. But of course it only seemed that way. They knew precisely where they were going. And when. The tangled web of booking changes was partly the result of the computer's habit of meticulously recording every move of eager ticket agents anxious to meet their customer's request for a particular connection.

The first suspicious booking, in the name of A. Singh, had been made on 16 June, a week before the tragedies. The passenger was holding a confirmed seat on CP Flight 003 to Tokyo leaving Vancouver on 22 June. From Tokyo's Narita Airport, he was booked on Air India Flight 301 to Bangkok. He never bought a ticket, however.

But on Wednesday 19 June, three days after the first Singh booking, a man called the CP Air reservations office seeking bookings for two other passengers. During a marathon conversation of over 40 minutes with agent Martine Donahue, he changed his mind several times before settling on a final agenda. Initially he asked for the same flight that had been booked for A. Singh, but he wanted to make a reservation for a man called Mohinderbell Singh. He did not mention the previous booking for A. Singh. He was out of luck, the agent informed him, for CP 003 was full on the 22nd. She suggested a seat on Flight 003 leaving on 21 June and a connection with Air India in Tokyo on 23 June. But the gap of two days between arrival and departure in Tokyo was not what the man wanted, so the agent offered him Flight 001, which was to leave Vancouver an hour after 003 left on Saturday the 22nd. However, both the agent and the customer wondered if that would leave the passenger with too short a time in Tokyo to connect with Air India. While they were talking, though, a seat became available on Flight 003 on Saturday, and the customer got the booking he had originally asked for.

During the same conversation, the caller also went to an extraordinary amount of trouble to book a flight out of Vancouver on the same Saturday that would connect with Air India Flight 182 before it left Canadian soil. He wanted a connection in Toronto for a man initially called Jaswand Singh (the booking was changed to the name M. Singh the next day). But CP 60 out of Vancouver on Saturday was full. Finally, he agreed to take Flight 96 out of Vancouver to Montreal's Dorval Airport. He was satisfied with that – for the moment.

Later on, after reservations agent Donahue left for home that night, the man called again. He didn't like the Dorval connection because it would mean transferring his bag from Dorval to the Mirabel Airport in Montreal, for it was from Mirabel that the international Air India flight would depart. But he was lucky this time; a seat had become available on CP 60 leaving from Vancouver on Saturday and the agent offered to put him on the waiting list for Air India Flight 182.

Going through the computer printouts of the history of the tickets, it became obvious to the Mounties that a man had taken great pains to connect two passengers with Air India flights leaving from Toronto and Tokyo. One flight – Air India 182 – was going east from Toronto, while the other – Air India 301 – was heading west to Tokyo from Vancouver. It was the first real break in terms of a connection between the two tragedies.

While the printouts gave them valuable information, they could not tell the Mounties how the man talked, what his accent was like, whether he sounded educated. But the agent who had spent 40 minutes on the phone with him could.

The clerk who could give the police those crucial answers was Martine Donahue. She would even remember that the man had a cultured voice, like that of Rajiv Gandhi, the Prime Minister of India. RCMP officers John Schneider and Bobby Sellinger would go over her story with her again and again. It was an amazing story. One interview started with Schneider asking her about the call on 19 June.

'In relation to the phone call that you received from a Mr Singh June 19, 1985 at the CP Air reservations office in Vancouver, will you please tell us exactly what you can remember about the conversation with this man,' Schneider asked.

'The gentleman who eventually called himself Mr Singh called [in] these reservations for two other people,' recalled Donahue, 'other than himself that is, one was to Bangkok and one was to Toronto; both had to link up with Air India.'

'At the time this fellow called, what exactly did he say to you?'

'Er, he just said, "I want to make bookings for two people, one going to Bangkok and one going to Toronto." So we worked on the, er, we started with Bangkok and then he specified that, um, it was to be on the Saturday on the ... I can't remember the date on Saturday on our flight to Tokyo to link up with the Air India flight going to Bangkok. I had a difficult time finding a connection so I suggested Hong Kong ... so we fiddled around a little bit and eventually found a seat on Air India which I booked and it was an excellent connection to our connection which ... required time at Narita Airport. After we completed that we discussed the fare,' said Donahue.

'Yes, what did he say in his own words?' Schneider probed.

'Well, first of all I had to get into contact with the tariff desk to get the fare and when I told him the fare there was no problem at all, there was no discussion at all on how high it was, because it *was* a high fare, no argument whatsoever. And then I said was the other person going, so we were looking for a flight to Toronto on Friday and there was nothing available.

At the time he called there was nothing available to Toronto either, so I suggested Sunday and we had flights available and he said, no, that would be too late, so I said, "You seem to have a deadline, what is it?" so he said, "It's to connect with the Air India flight" ...

'So I looked up the computer and found that the Air India flight left on Saturday from Toronto and made a stop in Montreal. Then I checked our flights to Montreal which happened to be available so I suggested that one, so I booked the other Mr Singh on that one ... I explained to him that he had to go, make his own way between Dorval Airport and Mirabel Airport, which is the international airport.'

'Were there any particular phrases or words spoken by the caller that seemed unique?'

'No, nothing outstanding; he spoke good English.'

'Is there anything else about the conversation that you can remember?'

'Nothing, no, the only thing is when he gave the names of Singh I said, er, I said, "You must be Sikhs with a name like that," and he sort of laughed and I said is your name Singh too and he said yes. So that's the only sort of personal contact there was in the whole conversation,' she replied.

'Do you think you could possibly describe his laughter?'

'Oh! Quiet. Like the rest of the conversation, it's not a peal of laughter. Um ... very pleasant voice, mature, er, poised voice with good command of English, a good choice of words ... simple but precise ...'

'Did he have any speech impediments?'

'No, not noticeably so. I could tell he was East India but not that accent that you could cut with a knife, probably somebody that spoke [English] fluently a long time.'

'Can you estimate the age of the caller?'

'Well, it's difficult to say, I'd say at least 40 anyway, 42 ...'

'Did he speak fast?'

'No, very forthright. About like you're speaking now.'

'And how about the depth of his voice? Was it a deep voice, high voice?'

'No, it was not a high voice, it was very poised, more of a deeper tone ... a nice voice.'

'Did Mr Singh mention to you who the tickets would be for?'

'No, he didn't. I knew from the start that it wasn't for him, so he must have said something to indicate other people ...' Donahue broke off, searching her memory.

'There's one flight to Bangkok, so did he, did he know that there was an Air India flight leaving Tokyo?' Schneider asked.

'He did, he did, because we had to link up and he also knew that CP Air connected with that flight, although my computer didn't give it to me, he knew it ...'

'Did you tell him that it would be cheaper to go through a local travel agency? Did he bring up anything to do with that?'

'No, the only mention of travel agency was when it came to picking up the tickets. I said where would you like to pick up your tickets? ... Most East Indians go through a friend or a travel agency ... I asked him if he was going to pick them up from a travel agency and he said, "Oh, no, we're not far from the airport, we'll go to the airport" ... oh, [I mean] he said "we're not far from you, from your office" ... but they went downtown finally.'

'So, he said they were not too far from the office, he would pick them up downtown?'

'He would pick them up from us, yes.'

'Okay, how about the other one?' Schneider continued.

'The other one ... I only booked Montreal. They wanted Toronto but there was nothing available,' said Donahue.

That particular interview was the third for her in less than a month after the Air India tragedy. Earlier, when Schneider asked her who she would compare the caller's voice to, Donahue had replied, 'Rajiv Gandhi. I've heard Gandhi talk quite often on television.'

In the first interview, she told Schneider that the conversation with the Singh who had called her probably lasted more than a half hour. She said she booked him on Flight 096 from Vancouver to Montreal because no flight was available to Toronto.

'What names did he give you to book the flights in?'

Schneider asked.

'I'm doubtful about the name he gave me. The caller spelled out a name of Mohinderbell Singh for the Vancouver-Tokyo flight, however, I don't know if he gave me his name or the passenger name. I definitely remember writing out the name Mohinderbell Singh on that reservation. I don't remember the name Jaswand Singh on the eastbound reservation. I don't remember what name he gave me for the eastbound flight.'

Later on that night, the caller who first talked to Martine Donahue telephoned again. This time, another agent picked up the phone. Mr Singh wanted to change his route for the eastbound flight from Vancouver. By this time, a seat had become available on CP Flight 60 leaving Vancouver on Saturday 22 June and connecting with Air India Flight 182 out of Toronto. This agent also tried to book a seat for Jaswand Singh on Flight 182 from Toronto, but he had to be put on a waiting list for the ill-fated flight.

Donahue told police that she thought the caller had made the change to Toronto from Montreal because he would have had to transfer his bag himself from Dorval to Mirabel. She told Schneider of a telephone number supposedly given as a means of contacting Mr Singh where he lived.

In an earlier statement on 24 June, one day after the tragedies, she told RCMP constable Bobby Sellinger and CP Air investigator Hank Martins that the man seemed determined to link up with Air India flights on both bookings. In one case, when she suggested he take CP Flight 001 from Vancouver, he declined because the flight left later than Flight 003 and would not be at Narita early enough to link up with Air India.

'I wait-listed him on the Royal Canadian Class [higher fare than economy class] on 003, that's the only one he wanted,' Donahue said. 'He didn't want 001, which I believe was available, and the reason he wanted [003] was that it linked up with Air India [in Tokyo].'

Sellinger was also told that the clue to whether a passenger had actually boarded a flight or not was to be found at the departure lounge, where the passenger presents his boarding pass, and where his flight coupon is pulled out before he goes aboard the airplane carrying the pass. If the passenger had not boarded, she said, the coupon would not be with CP Air personnel.

When the Mounties completed their interviews with Donahue, they were armed with some valuable information. They now knew that both tickets, one for CP Flight 003 to Tokyo to connect with Air India Flight 301 to Bangkok and the other for CP Flight 60 to Toronto to connect with Air India Flight 182, had been booked together. The bookings were for two people going in entirely different directions from the same departure point, Vancouver. It was also clear that they wanted to link up with Air India in both cases.

Why would anyone go to that much trouble and then not board his flight? That was the mystery which still had to be solved.

There was no further indication of what had happened to A. Singh, the man for whom a flight to Tokyo was booked initially on 16 June. The booking was never cancelled and no one by that name picked up a ticket. On Sunday 23 June, when the telex arrived from Air India's office in Tokyo asking what had happened to two passengers, the query was made regarding passengers A. Singh and L. Singh. Neither had showed up for Air India Flight 301 to Bangkok, said the telex to the Air India Vancouver office. A. Singh has remained a mystery, having left no trace except the Vancouver telephone number 324-7525 as his contact point. The number, belonging to the main Sikh temple in Vancouver, was given to Air India Vancouver by CP Air when an inquiry was made about where the passenger could be reached. That happened before CP Air security stepped in, pulling all information on the booking and ticket sales relating to the Singhs from the files.

When the telex from Air India Tokyo arrived in Vancouver, it raised a tantalizing question for investigators. The names of A. Singh and L. Singh: Could they be the two alleged terrorists Ammand Singh and Lal Singh hunted by the FBI in the US in connection with a plot to assassinate Prime Minister Rajiv Gandhi during a trip there prior to the Air India crash? Only days before the crash, the Mounties and the Canadian Security Intelligence Service had been alerted by the US Secret Service and the FBI that the men might be hiding either in Toronto or Vancouver, both of which have large Sikh communities.

Were they involved in the Air India disaster or did somebody want to make it look as though they were? That was the question the Mounties faced in the days following the crash. They began carrying pictures of the hunted pair as they knocked on doors in Vancouver to see if anyone could recognize Ammand Singh and Lal Singh, who had fled from an FBI dragnet in May, 1985.

Police thought it was quite possible that efforts to find the two in the Vancouver area had given local militants the idea of using the hunted pair's initials for the bookings to plant a false trail. Just about ten days before the crash, Canadian authorities and US Secret Service agents had visited some Punjab separatists to see if they recognized pictures of the two wanted men. The bookings were made after these pictures were shown around.

2

Tickets to Terror . . .

Mother Teresa's kind words have given comfort and hope to thousands of forsaken souls in the gutters of Calcutta. But there was no consoling Toronto. Not even the famed woman of peace could ease the grief suffered by the families of the victims of the horrendous Air India tragedy.

She tried, though, in her gentle, almost inaudible voice. It was Monday, 24 June. The Nobel Laureate was speaking in Toronto, her kind face etched with lines of age, the hands she held in front of her pitifully thin as she talked to the city which had suffered the biggest losses. Forgive, said the angel of mercy. Don't hold bitterness in your heart. But even the woman of God knew that day that a Jumbo jet doesn't just fall out of the sky like a stone without good reason. It wasn't an act of God – that was certain.

Her words were falling on the ears of people who were in a daze. What could you say to Yogesh Paliwal about his son Mukul? What could you say to Perviz Madon about why Sam hadn't reached Bombay for his birthday? Or to Mehnga Singh in Vancouver about why fate had snatched away his daughter Sukhwinder and grandchildren Parminder and Kuldip?

'We must pray that the families will be able to forgive,' she said quietly. 'They must forgive without allowing the bitterness or the anger to destroy the peace in their hearts.'

Mother Teresa sensed that there would be a need for

61

forgiveness. The media was already blaring around the world that sabotage was the likely cause of the Air India disaster. But each newspaper had a different story. Some said the bombs came from Toronto. Others said Vancouver. But the pundits all agreed that it was too much of a coincidence that the Narita blast and the Kanishka disaster had occurred within an hour of each other and that both planes involved had originated from Canadian soil.

But no matter how it had happened, explanations weren't going to help Yogesh Paliwal and Perviz Madon. Their anguish was too much to bear. Nothing made sense.

All they wanted to know at this time was whether they would be able to have one last look at the bodies of their loved ones. A team of pathologists headed by Cork doctor Cuimin Doyle were busy performing autopsies on the 131 victims found after the crash as relatives were arriving in the city en masse.

But in Cork, confusion reigned. The quiet sea-side resort had been turned into a morgue and was beseiged by crying relatives of the victims. Many of the families had been stranded in London, with Air India's offer to fly them to Cork now in utter disarray. The airline couldn't handle the logistics of such a tragedy. And even after the relatives arrived, many found that the bodies of their loved ones would not be released yet. All they could see were photographs for the moment.

The process of identification was slow and painful. A piece of familiar clothing, a document in a pocket, a driver's licence would give a clue. But the bodies were so brutally disfigured that mothers couldn't tell if they were looking at their own children. And relatives who did find the bodies of their loved ones were the lucky ones. Many did not, far more than half of the victims were still unaccounted for.

Perviz Madon finally recognized the body of her husband who had promised to be with her for his birthday. Sam had to complete his journey. She would take his body to Bombay. Yogesh Paliwal stood at the sea-side in Cork, staring at the horizon. He would not see the son who had kissed his feet and said goodbye. Shock and grief among the relatives had now turned to horror and anger. Who would blow innocent people out of the sky? Who could be so devious. So heartless?

Speculation does not solve the mystery of such a massive tragedy. Methodical and painstaking work had to be organized. Dozens of investigators had already begun to sift through tons of wreckage found floating around the coast of Ireland and brought ashore by rescue ships. More was washing ashore on the western coasts of Wales, Ireland and England, carried by strong ocean currents. A Canadian ship brought some of the wreckage to Halifax, Nova Scotia, while a Spanish ship took what it found to Madrid. On Monday, 24 June, a low-level search continued off the coast of Ireland involving merchant ships, Irish Navy vessels and US Air Force Chinook helicopters, but only one additional body was found. A majority of the bodies were those of females, some were infants, and the remainder were males.

In India, on 23 June, H.S. Khola, the 43-year-old Director of Air Safety in the office of the chief of civil aviation had already been given his orders to head the probe of the worst disaster ever to affect an Indian commercial jetliner. Khola faced a massive task. His only choice was to split the various investigators from the countries involved into small study groups, each with a specific area of investigation amongst the various aspects of the doomed flight.

Medical groups with experts like Dr Ian Hill, an aviation pathologist, would examine the bodies to try to determine the cause of the accident. Other experts, like C. D. Kohle, director of airworthiness for India's aviation authority, would team up with members of the Canadian Aviation Safety Board and the US National Transportation Safety Board to examine wreckage and eventually the Cockpit Voice recorder and the Digital Flight Data Recorder, the so-called black boxes which had yet to be recovered. The director of the Indian Central Bureau of Investigations, C. M. Sharma, was ordered to join the probe to look into the criminal aspects of the case and liaise with the RCMP and the Japanese police.

*

In Toronto, on Sunday, the RCMP had begun collecting duty rosters of anybody and everybody who had anything to do with the Air India flight while it was on the ground and when it departed. The list was endless. It included passenger agents from Air Canada who handled check-in; workers from an agency which handled the ground-transportation of baggage; staff from Burns Security who X-rayed and sniffed the bags with bomb detectors; catering staff and workers at the hotels where the crew stayed while in Toronto. Also to be interviewed were Air India agents D. Yodh, John D'Souza and Janul Abid. And the people who had associated with the flight crew while they were in Toronto, including Bhinder's friend Jagdev Singh Nijjar and the mysterious Sharma. The first interviews began in Toronto on Sunday.

Meanwhile, an ominous picture was already emerging in the Vancouver suburb of Richmond as the RCMP continued to glean clues from interviews with CP Air clerks. Although they had yet to obtain the full story of the two men who had checked in their bags but had not boarded their flights at Vancouver Airport, they had received a briefing from CP Air security about the incident. The first interview with Martine Donahue had provided an important clue.

The situation as the Mounties now understood it was that two men had checked in their bags for flights leaving out of Vancouver. Both of these CP Air flights connected with Air India flights and the men, both using the name Singh, had both failed to board their flights. Donahue's revelations provided the vital clue that the two men's tickets had been booked together by a single caller.

But Donahue had never seen the mystery-caller. The first face the Mounties could paste into their file was provided by a CP Air ticket clerk, Gerald Duncan. Never in his wildest nightmares had the veteran CP Air employee dreamed he'd be involved in a case of international intrigue. But now the Mounties were grilling him about a Sikh whose beard had fascinated him. They would even put him through hypnosis to probe the depths of his mind about the man who had picked up two tickets for terror.

Most passengers pre-book their flights. They come up to Duncan at the counter, identify themselves, and pay the fare. The computer prints out the ticket and he wishes them a nice journey. That's all there is to it, usually. But on Thursday, 20 June, things didn't work out that way.

Duncan was sitting behind a horse-shoe shaped ticket desk in CP Air's ticket office on Burrard and Georgia, the busiest corner in the city's business district. Selling tickets wasn't Duncan's regular job. His usual work station was the reservations office in the steel-and-glass Bental Centre where Donahue worked on reservations. But today he was substituting for an agent who was on leave.

He was doing the 10:00 a.m. to 6:00 p.m. shift when a tall, well-built man wearing a turban and a beard walked into the office and headed for Duncan's position. The man said he was there to pick up two tickets which had been booked earlier on the telephone. He now wanted to change the names in which the tickets had been booked. The booking for Mohinderbell Singh to Tokyo and Bangkok was now to be in the name of L. Singh; and that previously made for Jaswand Singh to Toronto and then New Delhi via Air India Flight 182 was changed the name of M. Singh. The man fished out a wad of money from his pocket. All cash. And with it was a piece of paper on which he had jotted down flight numbers and routes. The man said very little, but as Duncan worked on the tickets, he kept looking at the man's beard. It was neatly tucked under a black cloth net, which is worn to keep the uncut beard in shape.

There was nothing unusual about the sale, nor about the man's behaviour. Duncan would start thinking later about the fact that a man had paid cash for two tickets for passengers heading in different directions. While the man was there, though, Duncan got the distinct impression that the turbaned customer was a travel agent.

When RCMP officer Al Armstrong turned up to interview the CP Air agent, he carried copies of the two tickets Duncan had sold. The tickets told a remarkable story. The serial numbers on the two tickets were consecutive, meaning they had been printed out by the computer one after the other. The computer had also recorded that M. Singh's ticket was issued on 20 June at CP Air's downtown office in Vancouver with a confirmed status for CP Flight 60 to Toronto, departing Vancouver at 9:00 a.m. local time. The same ticket also showed an 'RQ' status, meaning wait-listed, for Air India Flight 181 from Toronto to Montreal departing 6:35 p.m. EDT (the Air India flight was designated Flight 182 when it departed from Montreal). M. Singh's ticket showed he was also wait-listed for the flight out of Montreal departing at 8:20 p.m. EDT. The total paid for the ticket was $1,697 in cash. The ticket number was 3522428.

L. Singh's computer-printed ticket number was 3522427. He had confirmed reservations for CP Flight 003 to Tokyo, departing Vancouver at 1:15 p.m. local time. His ticket showed that he was to depart Tokyo at 5:05 Tokyo time for Bangkok with Air India Flight 301. The payment, also in cash, for his ticket was $1,308.

Duncan, who has suffered anxiety and fear since the incident, described in detail to Constable Armstrong what had happened the day he sold the tickets, just two days before L. Singh and M. Singh were to fly. It was one of several interviews that he did for investigators over a few days' time.

'Could you please view the photocopies of airline tickets issued to L. and M. Singh on 20 June 1985 by yourself at the CP Air ticket office and tell me what you recall about these transactions?' asked Armstrong.

'I was working from 10:00 a.m. to 6:00 p.m. on 20 June 1985,' Duncan recalled. 'We can be sure it was early afternoon because it was only two hours till noon and I'm sure it was not right off the bat [that the man came into his office]. As far as I am concerned it was not one of the passengers that picked up the tickets. I got that impression because I thought he was an agent.

'I suppose I thought I thought that because the tickets were going both ways,' Duncan added. 'He kept referring to the people travelling as "they", so I presumed the tickets were for two other people.

'The man was as tall or taller than I am and I'm 5'11½". He

was in by himself – I'd describe him as 210-220 pounds, wearing a light gold-yellow or mustard colour turban. He spoke good English. He didn't have to repeat anything. Mind you, there is always a trace of an accent. He wore more casual Canadian-type clothes, nothing radical or no Nehru-type suit.'

'Do you recall any particulars on the dress?' asked the Mountie.

'No, it was just ordinary.'

'Do you recall anything else distinctive about his appearance, such as jewellery, scars, birthmarks?'

'No – he was not an ancient man. He wasn't grey – I'd say late 30s.'

'Do you recall him associating with anyone else in your office – employees or customers?' Armstrong probed.

'No – because he kept referring to the people who were destined to use the tickets as "they". I was sure he was by himself, usually if there was more than one, they'd both be at the counter.'

'Do you recall how he produced the money – from a wallet or in a roll, precounted?'

'When I got it, I think it was in a wad, folded in half. He may have known how much it was going to cost, because he had this information written on a piece of paper. He knew the flight numbers and routes,' Duncan replied.

'Did he keep that piece of paper?'

'Yes.'

'How long did this transaction take?'

'It must have been 10 to 15 minutes.'

'Do you recall why the first ticket was marked void?' asked Armstrong, referring to the mark put after the final destination, Bangkok, on L. Singh's ticket.

'See, the booking was already on the machine – I think. I don't remember making it [the booking]. That would mean someone phoned it in. See, the first ticket was booked "open return". I'm sure he knew what he wanted. It seems to me they were planning to come back, but they were planning on staying more than a year or they didn't know when they were returning so he changed it to a one-way ticket.'

'Was he concerned about the method in which the original flight had been booked?'

'Since the bookings were made, the people didn't know when they were coming back and that was it,' said Duncan about his conversation with the bearded man on the change from a return ticket to a one-way ticket. 'The guy was quite well-versed.'

'Did he comment about the short notice for booking or buying the tickets and the wait-list from Toronto?'

'No – probably because he looked to me like he knew what he was doing. He looked like an agent to me. There was never any discussion about alternative bookings.'

'Do you recall if he introduced himself when he approached your counter or identified himself at all?' Armstrong asked.

'No, he did not,' Duncan responded. 'Again, because he didn't do something like that, I presumed he was not one of the people travelling.' The ticket agent added that he didn't recall seeing any ride waiting for the man outside the office.

Duncan then described the turbaned man in great detail, as Armstrong continued to tape the conversation.

'This guy had a reasonably full face. Most of them look skinny and this guy was a big man. He seemed to have more meat on him than most of them do. Because I see so many I can't be positive about the description. He may have hair from his beard braided beneath his chin – it looked like cloth.'

'Could we just do a complete description, then?'

'Six feet, 210 to 220 pounds, full face, mustard turban, beard – possibly braided, casual western type clothes, spoke good English, no distinctive jewellery, carried list of flights, flight numbers and times and spoke as if tickets were for someone else. I'd say he was in his late 30s. Polite and maybe even soft spoken.'

The voice sounded like it was coming from a distant planet. It was soft but authoritative. It was gently nudging him into sleep. Probing. Taking him back in time. Back to the morning of 20 June. Driving to work, early morning coffee. Back in time, 16 July, 15 July . . . 22 June and then, slowly, to 20 June.

'And now I'd like you to begin orienting back in time,' the voice said. 'Just let your subconscious do this for you. Just as it has already slowed down your heartbeat. It has already lowered your blood pressure.'

Gerry Duncan was being taken back to the time when he came face to face with the turbaned man who bought two tickets from him. He was listening to the voice of the police hypnotist as the depths of his mind were being probed.

'It's almost like the development of a Polaroid,' said the hypnotist. 'When you take a Polaroid picture things don't seem too clear. Just say things that come to your mind as that gentleman approaches you at the ticket counter. What do you hear? What do you see? What do you feel?'

'Nothing,' replied Duncan.

Then, as the hypnotist probed his memory, slowly the picture appeared in Duncan's mind. Slowly, very slowly. Just like a Polaroid.

'The beard ... neat ... well trimmed,' he said. 'Thin ... parted ... and a mesh.'

'Yes, the colour of the beard?' the hypnotist prompted.

'It was dark ... maybe some grey, salt and pepper,' the CP Air agent recalled.

'Does he have a moustache?'

'I think so. It doesn't meet the beard ... It was a full face ... rough skin ... full lips ... clean turban, it was mustard.'

Then slowly, painfully, Duncan remembered the shirt, a grey shirt and a beige windbreaker. He also remembered a diamond ring. He remembered that the man didn't have enough cash on him so he changed the two-way ticket to Bangkok for L. Singh to one-way.

'Ah ... the money ... He went into his right pocket. He got the money from the right pocket. Hundreds and fifties ... folded.' Duncan remembered again that the man had the flight numbers on a piece of paper. 'It was just handwritten. It wasn't typed. It was just on a plain white paper ...'

The ticket clerk had thought he might as well kill two birds with one stone if he was going to be hypnotized. He wanted to stop smoking. So the hypnotist gave him suggestions about that too. And later he drew a composite drawing for the police.

At the time the tickets were picked up, Duncan had no idea nor even a vague feeling about what was to happen. But like other CP workers, he too had begun to worry after hearing the news of the disaster. From co-workers he'd heard that airport passenger agent Jeannie Adams had been involved in a fuss over baggage with an East Indian male. Out of curiosity, he called Adams to talk about what now sounded to him like an unusual ticket pickup. He also wondered if the man who had picked up the tickets resembled the man who had made a fuss with her about his bag.

Adams advised him not to tell the police that they had compared notes. But later she told police investigators what they had talked about anyway.

'This has been bothering me,' Duncan had said to Adams. '... The man who picked up the tickets had a turban and a beard.'

'Well, that's not my guy at all, 'cause I mean I know for sure mine didn't have a turban,' replied Adams. 'And he didn't have a beard.'

Relating the conversation to police, Adams added that Duncan had told her that when the man picked up the tickets he was fascinated with the man's beard net.

'He said he had a parting, he had this funny thing on and his hair was sticking through it, and I said, "Well, Gerry, East Indian men wear that." You know, that little netting that comes up the side?

'He said, "That really intrigued me ... I looked at that guy's beard and his silly little net more than I looked at his face." And when he said that a man had picked up both the Toronto Air India tickets ... it was a little strange that someone had picked up a ticket going to Delhi one way and another going the other way ...'

That was unusual, thought Adams, very unusual. It was indeed. It would soon be her turn to tell her story about the 'jerk' who had conned her into 'interlining' his bag at Vancouver Airport. A deadly bag, as all the evidence now indicated.

70

3

Mounties Make History

John Hoadley loves to fish. When he's not fishing for criminals he takes his boat out on his days off and throws in a line with a baited hook. Then he waits patiently for the fish to bite. The RCMP Inspector knows they will. Sooner or later. They always do. He can fool them all the time – whether he's fishing for crooks or out on the ocean. That's the story of his life. Bait the hook and wait. Then draw them in when they bite. His colleagues say suspects squirm uncomfortably under his cool blue gaze. He gives them the feeling he can count the loose change in their pockets.

'You show him a picture of five men standing together and you notice he's not looking at the faces,' said a police friend. 'He's busy looking at the car in the background and its licence plate.' The six-foot, blue-eyed Mountie with 14 years' experience of intelligence work has dealt with all varieties of law breakers, from petty thieves to PLO types.

On that Sunday afternoon when the tragic news of the Air India crash and the Narita blast was rocking the world, and when Ottawa RCMP brass started trying to locate John Hoadley, the head of operations of the Vancouver Integrated Intelligence Unit was out fishing.

The decision had already been made by RCMP Commissioner Robert Simmonds at headquarters in Ottawa that a massive, co-ordinated investigation had to be launched into the twin disasters. The small Richmond detachment just didn't have the capacity to carry out the largest investigation in Canadian history. The order from the top was that Hoadley, 49, should undertake the formation of the Mounties' largest-ever task force. The frantic messages finally caught up with him when he got home that day.

It was a formidable task. And Hoadley knew it. He would have to control the operations of more than 80 RCMP officers in Vancouver alone and co-ordinate their operations with up to 40 officers in Toronto and about 20 in Montreal. It had never been done before. The Crash Investigation Team would have well over 200 members country-wide, including dozens of support staff and clerical officers. But never before had Canada been the source of suspected aviation sabotage on such a disastrous scale. In fact nothing quite like this dual tragedy that took 331 lives had ever happened to the world at large, either.

The bosses had chosen Hoadley for good reasons. He knew the workings of the terrorist mind from many years of security and intelligence work, including his present stint with the Vancouver Integrated Intelligence Unit, a section of the National Criminal Intelligence Section. He was also thoroughly familiar with the structure of the Canadian intelligence community and the ways in which its work was divided between police and civilian branches.

It quickly became evident as the efficient but small Richmond detachment began probing CP Air files that Sunday that the scale of the investigation would have to be broader. Furthermore, it would have to be co-ordinated with other national jurisdictions, including Indian Intelligence and Japanese police. Careful co-ordination within Canada would be necessary too, for under Canadian law acts of terrorism fall under police jurisdiction only when a crime is actually committed. Before that point, investigating terrorist activity is the job of the Canadian Security Intelligence Service, but afterwards CSIS is obliged by law to hand over intelligence on the particular crime to the RCMP.

Hoadley's background made him particularly well qualified to deal with the problems of organization and co-ordination that were bound to arise. He had already accumulated more than a dozen years with the RCMP Security Service before it was disbanded and a civilian security agency was formed. Altogether, Hoadley had 30 years of police work behind him that Sunday as he undertook his biggest case ever.

The size of the task force meant that he would have to pull men away from other regular duties. Men from the commercial crime section, from the general investigation unit and even four Mounties from the Richmond detachment who were already busy sewing up some of the circumstantial evidence. The logistics of co-ordinating a team of some 140 officers were frightening. It would take at least three days before the team could be organized and working, and housing a team that large also meant clearing an entire floor at both RCMP headquarters in Vancouver and on Jarvis Street in Toronto, where that city's task force members would work under Inspector Seth Ginter, another veteran of the force.

Hoadley's other major problem was the initial lack of intelligence information, because the function of keeping an eye on elements perceived as a security threat had been the job of the Canadian Security Intelligence Service since the agency was formed three years earlier.

Hoadley didn't know it as he began plotting the formation of the task force on Sunday afternoon, but CSIS was about to give him an ace they were holding up their sleeve. The time had come for the men in the shadows to give the information to the Mounties, and they did so at a top-level meeting that same evening. The secret information, obtained through the tedious job of routine intelligence gathering, had already put CSIS on the track of a small group of Sikh militants. Their names were handed over to the RCMP, who quickly swung into action. Forensic work confirmed their suspicions that the small group of turbaned men were up to something sinister. But that fact was not enough to bring them to any solid conclusions. It was a lead, a very good lead, but much more work needed to be done.

This secret information, which had to do with explosives, meant that the Mounties would also have to be briefed on the nature of this particular Sikh fundamentalist group, which is small but deadly. The men recruited by Hoadley had little information about the activities of Sikh militant groups operating in the Vancouver area with connections in England, the USA, Germany and India. But CSIS knew all there was to know. They had become the experts in Sikh affairs and their advice is often sought by American authorities, according to intelligence officials in the United States. CSIS then briefed the RCMP on what this group stood for, what its estimated strength was and how it was financed. There was also a lot of helpful advice on how to operate within the Sikh community and deal with its particular sensitivities.

It was like a lesson in culture. Included in the crash course was the fact that you are supposed to cover your head when you enter a Sikh's home and remove your shoes. Just one of the little things the Mounties had to know as they came face-to-face for the first time with Sikh activists.

During the briefing, the Mounties learned CSIS had tons of information on Sikh militant movements. They had hundreds of names of associates of militant groups and information on those most likely to act out of hatred for India.

The Canadian security agency had first begun taking an interest in Sikh militants in the early 80s when a small, poorly-furnished office was opened at Kingsway in Vancouver. Inside sat a man called Surjan Singh Gill. He called himself the ambassador of the Republic of Khalistan, and, appropriately, behind his chair hung the flag of the new republic that he and his boss, Dr Jagjit Singh Chauhan of Reading, England, wanted to carve out of the Punjab State of India. Gill was busy issuing Khalistan currency and even passports of the Republic. At the time, though, he had few supporters among Canada's 200,000 Sikhs. Similarly, Chauhan was encountering the same problem in England and the United States, but he had made some friends in high places – including a right-wing US senator who had no love for India because of its friendship with the Soviets.

At the time, Chauhan was still in the early stages of preaching his gospel of a separate Sikh state. His most loyal supporter then, apart from Gill in Vancouver, was Washington resident Ganga Singh Dhillon, a millionaire who had already put out feelers for support from Pakistan for the

Punjab nationalist movement.

The security service operatives and this writer arrived at the office of the Republic of Khalistan two days apart. I had come as a reporter, to query Gill about his newly established but unrecognized diplomatic mission and about a statement by London-based rebel Dr Chauhan that he was giving military training to a group of his supporters in British Columbia.

Gill looked apprehensive. He wasn't sure how to explain his cause. It was his first encounter with the press and his voice sounded uncertain as he began. Later, though, he would show off the passports of his republic, its currency and even some stamps. But it became clear that the passports and stamps and Chauhan's statements about a military camp were of propaganda value only. There was no military camp. But the government of India was already fuming at Canada for allowing the office to open in Vancouver. Prime Minister Gandhi gave a piece of her mind to former Prime Minister Pierre Trudeau when they discussed the issue during an economic conference in Nairobi, Kenya in 1981.

Sitting in front of a picture of the Sikh Sovereign, Ranjit Singh, who had ruled the Punjab as an independent Kingdom towards the end of the 18th century, before the British Raj swallowed up the Punjab and the rest of India, Gill explained what his cause was. The government of Prime Minister Indira Gandhi, he said, was determined to wipe out the religious and cultural identity of the Sikhs, followers of a monotheistic faith galvanized by the tenth Guru Gobind Singh in 1699. Sikhs had to have a homeland of their own, he said, to safeguard the faith.

The Republic of Khalistan, the 'land of the pure', would provide a haven where Sikhs would no longer be drowned by the cultural influence of the Hindu majority of India, said Gill. Later, Gill introduced me to the President himself, a grey-bearded man with a smiling face. Dr Chauhan had paid me a surprise visit at my Vancouver apartment. He waved his plastic right hand and vowed to get his independent state within three years. By hook or by crook.

But that was only the beginning. Gill later split from

Chauhan's Khalistan Liberation Movement and teamed up with a Vancouver area Sikh fundamentalist named Talwinder Singh Parmar to form the Babar Khalsa, a movement which often cites the creation of the jewish state of Israel as a model for its aims.

That was back in 1981. At present, numerous Sikh fundamentalist groups have formed in western nations. Some are large and well organized, some are potentially dangerous but still in the organizational stage, and still others are lobbying groups who shun violence in the hope that they can convince western countries that they have a genuine grievance to air. Some of the most dangerous groups are, however, very small. Splinter groups can range in size from ten to 30 people, while the larger organizations in Canada boast membership in the thousands. In several countries, including Canada, the USA and Britain, they have become a top priority for security agencies like CSIS, the British Security Service (MI-5) and the FBI.

It was emphasized at the task force's briefing, to put things into perspective, that a vast majority of Sikhs in Canada would not endorse violence of any kind. They were too busy being Canadians to become embroiled in the politics of the Punjab. That even within the dissident movements, only a small core of people were capable of and willing to undertake an operation that smacked of PLO-type terrorism, and worse. What the police had to look for was a small cell of Sikh militants if they were to track down the culprits responsible for the terrorist bombs. A pocket of no more than ten to 12 people.

As Hoadley dispatched his men, first to nail down all the details they could from CP Air agents, then to dig up the information on criminals, it became quite evident to Hoadley that he was dealing with something that one was more accustomed to hear about in the Lebanon than in Vancouver: a genuine terrorist strike.

Hoadley found that he could not act as the administrator of the massive force as well as co-ordinate the criminal investigation and handle the deployment of personnel, hence

Mountie trouble-shooter Les Holmes was brought in. The superintendent, described by his co-workers as one of the best in the business, was a veteran of RCMP, having been with the force for nearly 30 years. An expert polygraphist and a veteran of homicide investigations, Holmes was now being given the job of administering the force of 80 officers and support teams based at the Vancouver headquarters. He was working on what tentatively appeared to be the largest murder investigation in Canadian history.

As Holmes took over administration, Hoadley took charge of operations. The two senior officers, Holmes and Hoadley, both with many years of experience in the workings of the criminal mind, and their massive team, would in the next few months carry out a mind-boggling operation that would conduct hundreds of interviews and generate tons of paper and reports. They hoped to make sure of one other thing too – to live up to the legend that the Mounties 'always get their man'.

4

Pesky Client

Somewhere tucked away in Jeannie Adams' memory was a vital clue. The RCMP had gone over and over the CP Air passenger clerk's story about the East Indian male with sparkling eyes who had appeared at her wicket on 22 June at Vancouver International Airport. First she had told her story to the RCMP officers from Richmond, then another official had arrived to go over her story to make sure nothing had been over-looked. After that she had gone over to police headquarters to make a composite drawing of the pesky passenger she wished she had never met. They had come back later to show her pictures of several possible suspects. Included with these were pictures of the infamous Sikh pair Lal Singh and Ammand Singh, the two fugitives from US law hunted by the Secret Service and the FBI.

She had shaken her head. No, none of these men looked familiar. And now the routine was to begin all over again. A tape recorder was whirring away as two Mounties asked the questions at RCMP headquarters on Heather and 37th Avenue in Vancouver. Other officers involved in the Crash Investigation Team were standing around.

Adams was nervous. Life had been miserable since that day the Singh had made a fuss with her to get his bag interlined directly onto Air India Flight 182 in Toronto. Often, in the middle of the night, she would run to the door. Would accomplices come back to get her for fear she could recognize the man? Would her name appear in newspaper headlines? Would she lose her job? Had she violated procedure by interlining a passenger's bag beyond the point for which he held a confirmed reservation? Her union had already asked her not to worry, they'd back her up if the company made any trouble. In any case, CP Air wasn't about to admit that a violation of procedure had occurred.

But now that tape recorder was making her nervous. She started to ramble, jumping from one place to another in her account. Constable Tom Armet, a towering Mountie who stands six foot two and weighs 190 pounds, noticed her discomfort. Relax, he said. Don't let the tape thing worry you. She was being asked to go back to the morning of Saturday, 22 June. Back to about 8:00 a.m. when the man had turned up in her lineup as she checked in passengers for CP Flight 60 to Toronto.

Passengers had begun crowding into the airport for weekend getaways and the first flights of the morning. At Air Canada and CP Air counters in the departure lounge, passengers were lining up for 9:00 a.m. flights to Toronto. It was like a madhouse that weekend, at least at Jeannie Adams' counter as she stood behind wicket 26. But it was just another Saturday in Adams' ten years behind the counter as she checked in the passengers one by one.

Wearing heels, standing about five-foot-six, in the usual grey dress of CP Air ground personnel, she surveyed the long queue of 30 to 40 people checking in at the counter. The time was shortly after 8:00 a.m. when she came face-to-face with a man with a 'cute' little face and 'sparkling eyes' whom she would also call a 'jerk' later. Life would never be the same again for her – she was about to become a pawn in a terrorist act which would send shockwaves around the world less than 24 hours later.

'It was a busy morning,' she recalled. 'And we had line-ups from the check-in counter to our ticket counter. It was extremely busy – I had maybe 30 people in my line-up or more – a very busy time of the morning.'

Usually check-ins are routine; at best they take only 30

seconds for passengers who are properly ticketed, Adams said. 'The passenger comes up, he has a ticket, the flight they are going on ... you ask them smoking or non-smoking and you check them in ...' That's how simple it is most of the time, and that's how it worked for her most of the time. But this time it was to be different. The passenger was a man with a very specific purpose. His ticket coupon bore the name Mr M. Singh. He was in his mid-30s, wore a grey suit and was about five foot seven. He spoke good English as he handed his one-way ticket to New Delhi to Adams.

'I remember the passenger because it was so busy and because he was taking up my time. He had a ticket to go to Toronto and then Delhi. This man, I checked his bag to Toronto. I put a Toronto tag on, I remember putting an orange Toronto tag on it. And I checked him in, gave him a seat and he wanted his bag checked to Delhi .... There must have been something wrong with his ticket because I looked at his reservation file. I went out of the check-in file (on her computer) and went into the reservation side – I remember his file clearly.

'He had Flight 60 to Toronto confirmed,' Adams recalled. 'Another flight, Toronto – Mirabel on the holding list. And another Air India flight, Montreal – Delhi ... and he wanted his bag checked to Delhi.

'And I said, "I can't do that, sir, because you're not confirmed on the flight." He said "Yes I am confirmed" and he said, "This is my ticket", – and I said, "Your ticket doesn't read that you're confirmed," and I said "I can't do it." He said, "But then I'll have to pick up my baggage and transfer it" and I said, "I realize that but we're not supposed to check your baggage through." But M. Singh wasn't about to accept that. He had to get that bag aboard by hook or by crook. He said, "I phoned Air India. I am confirmed on it."'

Adams, by this time slightly irritated by the man's insistence on getting his luggage sent through to India, tried to explain that he might be confirmed in Air India's computer but her computer said otherwise. His ticket only showed he had an 'RQ' – meaning requested reservation – but Singh stuck to his guns, maintaining that he didn't want the hassle of transferring his bag in Toronto. The line-up behind Singh was getting longer and longer. It was getting to a point where Adams was saying to herself, 'Come on, get on with it.'

Then the man said something else to her that made Adams more than just irritated.

'He said "Wait",' Adams said, recalling M. Singh's response. '"I'll get my brother for you."'

'The line-ups were busy. I thought, You gotta be crazy, and I said to him, "I don't have time to talk to your brother",' she explained to Armet and Constable John Hoffman. 'He started to move away from my line-up and leaving his baggage and his ticket. I said, "I don't have time to talk to your brother," so he came back and I said, "Okay, I'll check it through, but you have to check with Air India when you get to Toronto."'

It's a wonder Singh didn't jump for joy. The man standing behind him was by now paying attention to the fuss – he too suggested that Singh should check with Air India in Toronto.

'I remember ripping the tag off, and I remember thinking, You jerk, you're taking up my time . . .' said Adams. She said that she repeated at least five times to Singh that he should check with Air India in Toronto so they could take off his luggage in case there was no room on the flight. Then she put a pink 'interline tag' on the bag and wrote out the destinations on the tag: 'Vancouver – Toronto on CP, Toronto – Mirabel and Montreal – Delhi as the final destination.' She gave him seat 10b and wrote out his boarding pass.

'Can you just describe this Mister Singh, from the best of your memory?' Hoffman asked. 'Maybe just sorta think about it for a few seconds and then try and run a description by me, as much as you can recall, okay?'

'He was an East Indian gentleman,' she recalled. 'I feel if I had to describe him, he was probably someone born outside the country but had been westernized. Because he had longish, not longish but more curl – longer than yours – to the ears hair that was softly waved. It was in a western type style, not someone that you would consider who was out of the country. I think he was dressed in a suit – he didn't have a turban, I don't recall a beard. He had a kind of roundish-looking face – I remember thinking he was not a bad looking East Indian.

'Some of them are very foreign looking,' she added. 'I recall that he looked nice – he wasn't dressed, like, in Zeller's type clothes or with a plaid shirt and a – fairly nicely dressed . . . .'

'How about his language? Anything descriptive that might help us a bit?' Hoffman prompted.

'He had that East Indian dialect,' Adams went on, 'but it wasn't so heavy you can't understand it. Where some people, if they are born outside the country you have a hard time because they sound quite mumbly . . . he was definitely East Indian but not straight from the country.'

'He didn't have an earring in his ear or anything like that?' the Mountie wondered.

'No, he had a rounder face, sometimes they have very narrow, thin, gaunter looking faces or evil-looking faces, I remember thinking he had a rounder, kind of smiley-looking face, he was kind of . . . sparkly-eyed – pleasant.'

'Like he's a live sorta kinda guy,' added Hoffman helpfully. 'Not a dead sorta beat, I gotcha . . . .'

Adams then told the officers she felt that he knew the system, that he had travelled before and knew how the luggage check-in system worked. She added that he appeared very concerned about getting his bag aboard Air India. As she talked further with the Mounties it became apparent that Adams had worried later about having checked M. Singh's bag straight through to Air India.

'The only thing I feel very worried about now,' she said, 'is that there was no security check on Air India. So, if that is true, then I would . . . a straight flight . . . that, if I had called a, a, a supervisor, a supervisor could have over-ridden me too and said check him right through.'

'So what influenced you to check it through, do you recall?' Hoffman asked.

'Feeling that the passenger was right in that, if he had called Air India and had it confirmed, that if he had to pick up his baggage from CP Air in Toronto, which is very slow at uh, baggage output, and having to pick it up and go upstairs and check in with Air India he might miss his flight,' Adams replied.

'Now, what about his brother, the brother he said he would get to prove he had called Air India?' Hoffman asked.

'He was going to leave my line-up to get his brother. And I thought, You're crazy, I don't have time to talk to your brother,' Adams replied.

'Did you ever see his brother?'

'No, but he was leaving his baggage and his tickets to go get his brother,' she said. 'So, if I had called his bluff, he'd have had to get someone.'

Adams recalled that the man had only one bag. She remembered that because she had ripped off only one tag from it. The one she had originally marked for Toronto, before substituting the new one, marked Toronto–New Delhi. The bag probably weighed about 50 pounds, she thought, and her best recollection was that it was burgundy in colour.

Another thing she remembered in the conversation with police was that the pushy Singh had reminded her that he was travelling business class.

'He said, "I'm paying full fare to go to India. That's why I paid full fare, so that I could get my bags checked through" ... I recall that.' Adams added that paying the full fare, business class, sometimes gets you better treatment than the excursion fare.

Adams – who, incidentally, is married to Bob Adams who also works at the Vancouver Airport – told the police officers she was relieved when Singh finally left with his boarding pass. He had taken up more than five minutes of her time making a fuss about his bag, although he was no more pushy than other passengers sometimes are on other routes. Sighing with relief that he had gone, she put on a smile for the man who had been waiting patiently behind Singh all this time. She apologized to the middle-aged man. He said it was okay, he understood.

Later she recalled how she started putting two and two together, after Air India Flight 182 had plunged into the Atlantic. Tom Armet asked her if she recalled discussing M. Singh with any of her co-workers.

'Only about two or three days later, when I thought of it then, I thought, I didn't realize the London plane that went down was carrying on to Delhi and I said, I remember having a hassle with one passenger,' she replied. 'I said gee, I felt bad now. Maybe he was on that plane . . . then I put two and two together.'

Adams was in for another surprise as she talked to the RCMP. As a matter of fact it came as a shock when a police artist gently told her that she also might have been the one who checked in L. Singh for his flight to Tokyo via CP Air Flight 003 connecting with Air India Flight 301 to Bangkok. L. Singh's reservations were in order and, unlike M. Singh, he was confirmed on both flights. But he didn't board the flight for Tokyo.

Adams is still wondering today if he was the brother M. Singh referred to when he told her he would get his brother to confirm that he had called Air India and they had okayed his trip from Toronto to New Delhi.

'I thought [I had done] well, after I did the composite drawing with Constable Blair, [and] he said okay and how do you rate them? I said six to seven [on a scale of ten]. He said, "Now, what do you remember about the other guy?" And I said, "What other guy?" [He said] just the guy who went Vancouver, Tokyo . . . And I said, "Well, why should I remember him?" And he said, "Because you checked him in too" and I just went . . . I said, "You gotta be kidding!" . . .

'And I said I didn't realize that and I said, "Oh, rats!" And we kinda half-joked about it and I said, "Darn it, I'm gonna be done out of a job here," and he said, "We thought . . . we didn't think we would tell you before you did the composite drawing 'cause it might be on your mind."'

Adams' ticket agent code, D-2, was on L. Singh's ticket as well as M. Singh's.

RCMP did a thorough check of whether an M. Singh had boarded the CP Flight 60 from Vancouver. It became evident that he hadn't. On Air India Flight 182, there was another man with the same initial and surname. He was Mukhtiar Singh, an Indian tourist who had bought his tickets in India

84

and was returning home. He was 58 years old and his body was found and identified in Cork.

The picture was complete now. The circumstantial evidence was in. At a meeting at RCMP headquarters, police took stock of their findings so far. It was a startling picture that could be explained only if Air India Flight 182 had been downed by a bomb.

As they considered what to do next, this is the point-by-point summary that members of the task force were presented with:

*One man had made two bookings over the telephone for passengers travelling in different directions from Vancouver. Jaswand Singh's booking was for CP Flight 60 to Toronto and he was wait-listed for Air India 182 out of Toronto for a flight to New Delhi.

*Mohinderbell Singh was booked to fly CP Air Flight 003 to Tokyo and from there he was to fly to Bangkok on Air India. He held confirmed seats on both routes. The day the bookings were made was Wednesday, 19 June 1985.

*In the afternoon of 20 June, one man, a bearded and turbaned customer, had come into the CP Air office and changed the names of the passengers. The man flying Vancouver – Toronto – New Delhi was now to be called M. Singh. The man flying Vancouver – Tokyo – Bangkok was renamed L. Singh.

*While most Indo-Canadians book their flights from local travel agents who offer discount fares, the bearded man had chosen to pick the tickets up directly from CP Air, most likely because the chances of recognition in a busy office manned mostly by white workers would be slim. The payment was cash and money was no object.

*Both men flying in different directions were to connect with Air India flights. Both checked in their baggage at Vancouver airport on the same day, 22 June. Neither boarded their flights.

*CP 003, on which L. Singh was the only passenger who had checked in his bag but did not board, had carried a bomb to Tokyo.

*Air India Flight 182, carrying the bag which M. Singh

had made a fuss to get interlined with this flight, exploded over the Atlantic within an hour of the Narita blast.

The Mounties were now convinced that only believers in lightning striking twice could dispute the fact that the two disasters were linked and that there had been a bomb aboard Air India Flight 182. They had also concluded that whoever had put those bombs on board the planes was deliberately targeting Air India flights in different parts of the world, and hoping for near-simultaneous explosions. Few people could argue with that logic.

Absolute confirmation was still needed, however, that M. Singh's bag did in fact get aboard Air India 182. That answer would be found in Toronto. But first, another piece of the jigsaw had to be put in its place in Vancouver.

5

Strange Call

The man was worried about getting his bag on the Air India
Flight. Determined. It was the only thought in his mind. And
he made it apparent to Canadian Pacific Airlines reservations
clerk Aziz Premji just two hours before he went and argued
with Jeannie Adams at Vancouver International Airport. He
wasn't worried about getting himself on the flight. Just his
bag. He didn't care about alternative routes. No, he just
wanted Air India 182 out of Toronto.

It was 6:30 a.m. on 22 June, Premji, an East African
immigrant to Canada, was sitting at the reservation computer
in Bental Centre when the telephone rang. It was M. Singh,
or Manjit Singh as he called himself later. Singh picked up
the slight hint of an Indo-African accent in Premji's voice and
began speaking to him in Hindi, laced with Punjabi. Singh
even asked if Premji was from India. No, said Premji, he was
from Nairobi.

Premji thought nothing further of this telephone call until
Monday 24 June when Mounties Bobby Sellinger and John
Schneider walked in to talk to him. That was the first time.
Then Schneider came in again, this time accompanied by
multi-lingual Mountie Sandy Sandhu. They went over his
amazing story again and again. The following is one extract
from their interviews with him.

'I got a call from Mr Singh who wanted to know if his flight to

Delhi was confirmed,' Premji told the Mounties. 'So I asked him for the flight number and he said he was taking a flight from Toronto. Then I asked him if he was going on CP Air to Toronto and he said he was – so I looked for the flight number myself because he didn't know what the flight number was.

'Looking at the PNR, I found that he [M. Singh] was confirmed on CP Flight 60 leaving Vancouver at 9:00 a.m. That would have brought him to Toronto at around 4:10 p.m. [Toronto time] then the flight out of Toronto to Mirabel Airport on Air India. I think flight 181 was wait-listed, flight 182 from Mirabel to Delhi was wait-listed and I recall he was on a one-way ticket full-fare so I offered him an alternative route via the Pacific by Tokyo and he said no.

'He wanted the same flight because he had friends going from Winnipeg to Toronto connecting on the flight, so I asked him if he would give me the names of his friends so we could maybe figure if they are confirmed or not.'

He did not know it at the time but Premji had just called the man's bluff. The sly customer was evasive in his answer, however.

'No, that's okay,' was the reply from the Singh.

Then came the question that showed what M. Singh was all about:.

'He wanted to know if he could send his baggage right through to Delhi and I said no, he can't do that because his flights were wait-listed. So what if he could give his baggage tags to his friends who were coming from Winnipeg, [he asked,] and I said no, you can't do that, and he said he would try on the flight anyway and see if he can get on a flight out of Toronto.

'During the conversation he recognized my accent,' Premji added, 'then he asked me if I spoke Hindi and I said yes.'

'How long did he speak to you in English before he spoke to you in Hindi?' the policemen wanted to know.

'About a minute or so,' Premji replied.

'And how long did he speak to you in Hindi?'

'I think the whole conversation was about five minutes, nothing more than that, so, five minutes.'

'Can you please describe his voice for us?'

'He didn't speak very loudly, it was a soft voice, I would say.'

'Did he have a strong accent?'

'Well, he had an accent when he was speaking in English, but when he spoke in Hindi he spoke more Punjabi than Hindi words or, you know, maybe I got that impression he spoke more Punjabi.

'He was aware of what he was talking about, like he mentioned baggage tags and not many [Indian] people know what baggage tags are, and that if he could give baggage tags to his friends ...' Premji added.

'Did he have knowledge about Canada like did he know where he was going, Mirabel, Dorval, did he know the difference?'

'No, I didn't get into that really, the only thing I told him [was that] his flight was wait-listed from Toronto to Montreal, he didn't go into the airports.'

'Did you find it odd for someone to travel to Bombay through Toronto, because it's much easier to travel from Vancouver to Tokyo to Bombay?'

'Not necessarily, because they might have got a better fare on Air India out of Toronto, so there's nothing unusual about that.'

'Again, getting back to the language, did he appear to be new in Canada, like, say, in the past five or six years, or did he appear to be Canadian-born from East Indian descent?'

'No, I don't think he was Canadian born,' was Premji's opinion.

In an earlier statement to police, Premji said he was quite surprised when M. Singh asked about giving his baggage tags to someone else to pick up. He also said the caller sounded educated and like someone who had travelled before. To Premji, the caller sounded about 40 years old.

'Did you hear any background noises, music, radio, possibly noises from outside a phone booth?' the investigators asked Premji.

'No, it was a local call, you can tell on the phone and I'm pretty sure it was from a house or building; I didn't hear any unusual or noticeable noises in the background.'

One more piece had been added to the structure of circumstantial evidence of a bomb on Air India 182. The Mounties had done well so far. Now it was time to ask their Toronto colleagues what they'd found out from the probe in the East.

6

Where's The Bag?

RCMP headquarters at 225 Jarvis Street in Toronto was a beehive of activity. The list of people to be interviewed about the fate of Air India 182 was mind-boggling.

The list included a number of Air India employees including passenger agent Divyang Yodh and security man John D'Souza; Air Canada technicians who loaded the extra engine and prepared the aircraft for takeoff; employees of Burns Security who were assigned to Kanishka; Air Canada passenger agents who checked in the passengers; ground baggage handlers; four RCMP officers who were sitting under the plane keeping a watchful eye on it; cabin cleaners and baggage container loaders; anybody and everybody who had even a minor role to play while the aircraft was on the ground in Toronto. And even Jagdev Singh Nijjar, friend of co-pilot Bhinder, and the mysterious Sharma who had given them both a hard time.

There were so many questions to be answered. Did all the passengers who checked in board the aircraft when it left Toronto? What is the procedure for movement of baggage from the check-in area to the final point of loading? How many passengers were brought to Toronto from outside the city by connecting flights? Who did the crew of Kanishka associate with in their last days in the city? If there was an explosive device in a passenger's bag, why wasn't it detected

91

by X-ray?

But the biggest question of all was: Did M. Singh's bag definitely get aboard the doomed flight? Did anyone remember what it looked like; what colour it was?

The Mounties, under the direction of Staff Sergeant Mike Atkinson of the national crime intelligence unit, began charting the movement of bags at Toronto's Lester B. Pearson Airport on the day the flight left for its date with destruction. The 20-year-veteran of the RCMP knew his first task would be to prepare a video film of the operations of the airport. The purpose would be to determine exactly how luggage transfers were made at Toronto Airport, which is split into two terminals.

Terminal One handles most of the domestic traffic, including Canadian Pacific and the international airlines for which it is a handling agent. Terminal Two is leased by Air Canada and serves airlines that it handles, including Air India and other overseas airlines. The job of picking up bags from interlined passengers arriving at Terminal One rests with a company called Consolidated Aviation Fueling and Services Ltd (CAFAS). Atkinson's men would find that the CAFAS driver would start his run from the Air Canada sorting area and deliver any baggage destined for domestic runs to Terminal One. The driver would then return to Air Canada's sorting area with any baggage to be interlined to Air Canada and flights for which it is the handling agent. The CAFAS driver maintains a record of baggage picked up from the terminals on a time sheet for purposes of billing the airlines.

CP Flight 60, carrying M. Singh's bag, arrived at Toronto's Terminal One at about 4:10 p.m. Toronto time and docked at gate 51 with at least a dozen passengers who were connecting with other flights to continue their journey. The CAFAS driver's records for 22 June show that he made three trips to CP 60 to pick up baggage to be transferred to Terminal Two, where Air Canada was handling the Air India bags. The three trips occurred between 5:00 p.m. and 6:30. The Mounties believe M. Singh's bag would have been picked up by the CAFAS driver on his first trip, which was almost an hour after CP 60 had arrived.

Just to make sure, the RCMP contacted each of the passengers who had come in on CP Flight 60 from Vancouver and connected with other flights to continue their journey onward from Toronto. All of them said they had received their bags at their final destinations. There was no reason, then, to believe that M. Singh's bag had not made it to the luggage sorting area from which Air India 182 luggage would be taken to the flight.

The RCMP also noted that Air India would not even be aware of the fact that M. Singh's bag was being loaded onto their flight, because the bag, carrying an interliner tag for New Delhi, would automatically go to the sorting area for a security check along with other transit luggage destined for Flight 182. The Mounties would find that Air India, which had already pointed out to the Canadian government in the spring that it felt India's commercial installations in Canada such as The State Bank of India and its own operations were a potential target for terrorists, were using a novel system to make sure all passengers who checked in actually boarded the aircraft.

The check-in of passengers had begun early on 22 June. Initially two counters had been opened and subsequently the number was increased to six, serving both economy and first class passengers. Passengers were being checked in by Air Canada personnel because the Canadian airline is the handling agent for Air India. According to the system of security recommended by the Indian airline, the Air Canada personnel would write a number on the boarding card of each passenger which would subsequently be compared with a control sheet at the time of boarding to make sure that all passengers who had checked in had boarded. And that's exactly how the system had worked that day.

M. Singh did not check in with Air India, so his name would not appear on the security control sheet designed to make sure that all passengers who had checked in were aboard. A thorough check of the entire terminal showed no sign of any bag left behind. There was no doubt M. Singh's bag had gone aboard Air India, a bag Jeannie Adams had recalled as being either grey or burgundy in colour.

*

93

On Monday 24 June, RCMP officer K.T. Kervin talked to Burns Security staff member Naseem Nanji, who had heard a beep while a fellow employee was using a bomb-sniffer on a bag she thought was maroon in colour. Later, she was interviewed for a second time, along with fellow Burns employees, by RCMP officer B.P. Thomas.

Nanji, 35, gave this statement to the RCMP Air India Task Force:

'I joined Burns International Security Services on or about the 24th of June 1984. I was working full-time till January 11, 1985. On the 19th of January 1985 I started working the Air India flights on Saturdays only. My shift was from two o'clock to six o'clock in the afternoon. Normally I worked near the Canada customs area, watching the baggage belt area. My responsibility was to make sure no passengers put their luggage on the belt after being checked through. As I speak Hindi, I helped Air Canada staff if needed.

'I think on the 1st of June 1985 I was switched and worked with the X-ray scanner. That day I was trained by Burns employee Stanley Noble,' she revealed. 'He gave me the impression he didn't know everything about the shapes on the screen, but explained how to look for weapons such as guns and knives.'

Nanji also told the police how little real training in security work she had received from Burns: 'I was shown how to move the X-ray unit into place, plug it in and how to open the box on top to turn on the switch. Then I was shown how to pass baggage through the scanner and get a proper picture. I never attended any other training course to learn how to operate the X-ray scanner. The only course I got from Burns was first aid and CPR! (Cardio-Pulmonary Resuscitation)

'On my second time using the X-ray scanner, on Saturday, June 8, it broke down and we used the sniffer that day to check the luggage for the [Air India] flight,' Nanji said. She worked the X-ray machine by herself on 15 June, with colleague Jim Post and another man helping her put the luggage on the belt of the scanner.

'By the way, I was told if I saw a weapon in the X-ray picture I was to test the baggage with the sniffer,' she said. 'I was only to check with the sniffer if I saw a gun.'

Talking about her first day with the scanner, when Stanley Noble showed her what to look for, Nanji said she had seen the shape of a gun, but upon examination it turned out to be a toy gun.

'I didn't receive any instructions for how to look for a bomb,' she added. 'I was told to look for funny wiring or connections. If a stereo receiver went through, it showed up on the screen. We often had stereo receivers go through in the suitcases. I think they were hidden in the suitcases so they didn't have to pay duty over there.'

It was on 22 June while she was working on the baggage, putting it through the X-ray scanner that she had only used three times, that the machine packed up. The picture disappeared from the screen. The time was around 4:45 p.m., she said.

Air India security man John D'Souza came along then and tried to fix the scanner but couldn't. So he instructed Nanji and Jim Post to use the PD-4 explosive sniffer.

The CAFAS driver had just brought a cart of transit luggage from passengers who were coming in from other cities to connect with Air India 182. The time was around 5:00 p.m.

'Between 5:15 and six o'clock I heard the sniffer beep when it checked a bag. Post was checking around one bag's zipper when it beeped,' said Nanji. 'This bag was soft sided and had a zipper going all around it,' she recalled. 'I believe it was maroon in colour.

'James Post did a second check, and it beeped low in volume when it passed near the zipper's lock,' she added. 'But the beeper wasn't making a long whistling sound like it had when John the Air India man demonstrated the sniffer to us.

'I never told Air India about these beeps because no one told us to call them if the sniffer [only] gave a short beep,' she admitted.

Furthermore, Nanji said, during the X-ray checks, while the machine was working, she had seen more stereo radios inside some of the suitcases. She also saw a television set that had been checked in and cardboard boxes full of nappies.

The X-ray machine was checked the next day, but no fault was found. Technicians told the Mounties that moving the machine back and forth may have caused a temporary blackout of the screen. There was no reason at all to believe that the machine had been tampered with. Technicians said they had fixed the device several times in the past because of problems caused by movement. The scanner, the Mounties were told, is designed to be put in one place and left there.

Next the Mounties interviewed Jim Post, the man who was using the hand-held explosive sniffer that Nanji and other workers said had given off a beep while he was checking a bag. But Post didn't seem to be very happy about being interviewed. In fact, he wasn't even happy that he had been assigned to work at the Airport.

Corporal B.P. Thomas was asking the questions. 'How long have you worked for Burns security?' he began.

'Eight months. January 10th, 1985 I started.'

'How often had you checked at the airport for Burns Security?'

'I don't know the exact dates. It's on the time cards,' replied Post.

'Have you received [the] transport Canada passenger screening course?' asked Thomas.

'No. I was taken out and dumped in and told to do the job of checking baggage,' Post told him.

'What training have you received from Burns Security?' the Mountie asked.

'None at all, except the eight films for [the] Burns Security Orientation program. I got eighty-six per cent on those tests. I haven't had any training at the airport.'

'When the X-ray unit broke down, who instructed you to use the sniffer device?'

'I didn't have to be instructed,' replied Post. 'Common sense told me to use it.'

Asked what the sniffer was supposed to do should it detect an explosive, Post said he had been told it would make a loud whistling noise.

'On Saturday, 22 June 1985, did the sniffer ever make a

loud scream while you checked any bags?'

'No. The only time it made a beep was when it was turned on or off. I would turn on the sniffer when a bag or baggage came. After the bag or baggage was checked I turned it off till the next baggage came. Each time it would beep as I turned it on and off.'

'Did any Air India personnel show you how the sniffer worked that day?'

'Just this fellow with the handlebar moustache and an air head,' said Post, referring to John D'Souza, Air India security officer. 'He came up to me while I was using the sniffer. He acted like he had all the answers. He said, "You use the sniffer this way." I handed it to him.

'He lit a match and held it about an inch away and it made a loud piercing scream. He told me to go around the edge of the bag. He didn't tell me to push on the bag [to expel air from inside so the bomb detector could sniff it]. Even if this incident hadn't happened I was going to tell Personnel I would not be going to the airport again.'

Post declined to sign the statement he had made to the Mounties.

He had also insisted on the presence of a friend while making the statement and on being allowed to read it over and correct it by himself.

During the loading of containers filled with luggage aboard the plane, conversations between the loaders, workers for Mega International who are cargo handlers for Air India, and ramp personnel are recorded. From these conversations the Mounties were able to draw exact diagrams showing which containers carried which bags when the plane finally left Montreal. They were also able to draw charts of the forward luggage compartment with its 16 containers which held 338 passenger bags.

Two containers close to the sensitive electronic bay of the aircraft contained bags bound for New Delhi from Toronto. M. Singh's bag could have been in either of these two containers, since Jeannie Adams had marked it as destined for New Delhi. Three of the forward luggage containers were empty, while one held diplomatic bags and valuables weighing less than 20 kilograms. A container towards the rear of the front luggage hold carried fan blades and parts of the extra engine in wooden boxes.

The aft cargo compartment carried four pallets towards the front. They were loaded mostly with spare parts of the engine as well. The two rearmost containers also carried Toronto-Delhi baggage. The bulk cargo compartment, which is the smallest space on the aircraft's sloping back section, held 27 bags bound for Delhi from Toronto.

Toronto Mounties also made the following notes from their extensive interviews:
* The crew of Kanishka were not subjected to security frisking, although their bags were screened.
* Two diplomatic bags from Vancouver and another from Montreal, because of diplomatic protocol, had gone aboard without being X-rayed or checked.
* Captain Bhinder had been seen carrying parcels wrapped in brown paper aboard the aircraft.

Intelligence reports suggested that Bhinder had associated with Jagdev Singh Nijjar, whose brother Balbir Singh Nijjar is one of the most ardent supporters of the Sikh separatist movement in India. Balbir Singh has recently been posted to Equador as ambassador of the unborn Republic of Khalistan.

Nijjar later told this writer that police had questioned him about his friendship with Bhinder.

'At no time did I give Bhinder any parcels,' he told me. Intelligence sources have confirmed that through exhaustive inquiries they reached the conclusion that Nijjar's friendship with the co-pilot was genuine, and they have no suspicions of Bhinder having carried any booby-trapped parcels into the cockpit. Canadian authorities also believe that Bhinder went shopping while he was in Toronto and carried his purchases wrapped in brown paper. Ground personnel who saw him boarding the plane said he carried a parcel about the size of a shoe box along with his flight briefcase.

Exhaustive police inquiries in Toronto also ruled out any booby-trapping of the extra engine, the doors of the Jumbo jet or the containers. But there was one curious incident that was reported to the RCMP by Burns Security officer Jack Prosser. He was working on crowd control when at around 5:30 two youths came up to him and said they wanted their mother off the flight. Prosser said one of the youths told him he had had a premonition about the flight and his mother should be allowed to get off. The security man escorted the two youths to an Air India representative, but they didn't persist in their effort to get their mother off the plane. They didn't say what the nature of their premonition was.

It was Sharma who told the Mounties what he had been up to the evening before the doomed flight when he entered Bhinder's room while the co-pilot was with his friend Nijjar. They learned that the 26-year-old man was an insurance sales-man. He was trying to deliver a message to a London, England resident named Tahir Ali, a former agent of the Indian Intelligence agency called the Central Bureau of Investi-gations. London police interviewed Tahir Ali and he admitted knowing Sharma. But he said he didn't know what message Sharma was talking about. RCMP believe Sharma was using the supposed message as an excuse to meet Captain Narendra, who was refusing to see him.

Vancouver RCMP officer S.R. Miller went to Toronto to track down the CP Air 60 passenger who had patiently waited behind M. Singh as he argued with Jeannie Adams about his bag. This passenger turned out to be a former manager of Toronto Airport, and he remembered the East Indian male who carried a briefcase and wore a grey suit.

'When the East Indian male reached the check-in agent, there seemed to be a considerable discussion between him and the agent,' recalled the man, saying he had travelled on his birthday from Vancouver to Toronto on 22 June. He didn't know then, of course, that he had travelled on a plane that had carried a bomb!

'The gist of the conversation which I recall was to the effect that she could only check his bag so far as Toronto,' the man said about the incident in Vancouver. 'He was only wait-listed beyond Toronto.'

The former airport manager also told Miller, 'I did not observe any jewellery. As I stood behind him his hair was well coiffeured. He [looked like a] neatly attired, well-turned-out businessman.'

A businessman, indeed. His only business that day, it appears, was to check in a bag bomb for Air India 182.

Another man the RCMP interviewed in Toronto was Otto Von Staffeldt, an Air Canada passenger agent who had worked on the first and business class counter. Here he had to check in the crew as well and take charge of unaccompanied children being deposited at his counter by their parents.

Von Staffeldt was almost reduced to tears when he recalled the children on Air India 182. How he had assured their parents that the kids were in good hands. How he had taken them to the flight and handed them over to the hostesses.

'I spoke to all the parents [of seven unaccompanied children],' he said. 'I escorted them to the departure gate 55 minutes before departure. I then took them all on board to their seats, turning each one over to the flight attendant.

'It has been my most devastating experience with Air Canada to date, to hear of the disaster the next morning,' said the agent about the tragedy. 'Remembering the children . . . I personally took by the hand on board the aircraft and the trusting parents I assured [that their children] were in good hands . . .'

He also recalled seeing co-pilot Bhinder with a brown parcel.

'Bhinder did have brown wrapped packages in his buggy. I don't recall how many packages he had,' the agent said.

'The packages that S.S. Bhinder carried on board, can you describe them?' asked Corporal Thomas.

'They were wrapped in brown paper. They were sitting on top of his flight briefcase. I can't say for a hundred per cent but there were one or two.'

'Did they look heavy or light?'

'I wouldn't know, because they were on Bhinder's buggy,' replied the Air Canada agent, who obviously had been devastated by the deaths of the children he had taken to their seats.

7

The Hunt For Black Boxes

As the Mounties, the Japanese police and the Indian Central
Bureau of Intelligence pursued the criminal aspects of the
twin tragedies, the foremost task facing accident investigators
in the days following the Air India crash was to find the all-
important 'black boxes' from the 747. At the site of the crash,
the depth of water was approximately 6,700 feet. Air India
investigator H.S. Khola would call it a 'very difficult and
challenging job'. But it had to be done if any sense was to be
made out of what had happened to the aircraft.

Time was running out day by day. The black boxes have
enough power to transmit signals for approximately 30 days.
Ten days after the plane had gone down, there was still no
sign of a signal from the black boxes.

The boxes can provide vital clues. The Cockpit Voice
Recorder contains an endless magnetic tape which records
conversations from several mikes in the cockpit. It erases the
tape and begins recording all over again every 30 minutes.
What were the crew saying at the time of the crash? What was
their reaction to the emergency? Did they have time to say
anything at all? Only the voice recorder could answer these
questions.

The Digital Flight Data Recorder was even more
important. It keeps tabs on 52 different aspects of the flight,
including aircraft heading, altitude, position of landing gears,
the thrusts of the engines and the position of the rudder,

among other things.

According to an official report by Khola, three ships were handed the task of first locating the boxes and then bringing them to the surface. The *Guardline Locator*, a ship provided by the accident investigation authority of the United Kingdom, *Le Aoife*, an Irish naval vessel, and the *Leon Thevenin*, a French cable-laying ship chartered by the Indian Government, finally began picking up pings from the bottom of the ocean on 4 July. The pings were first picked up by the *Guardline Locator*. It could hear two separate sources of sounds. It gave the *Leon Thevenin* the location of the sound sources.

The French ship began searching the general area but did not have any success until 9 July. Then, using a cable and its underwater mini-submarine the *Scarab*, the vessel plucked one of the boxes from the deep. It was the Cockpit Voice Recorder. The next day, sonar picked up signals in the same area. Again, the miracle-working *Scarab* went to the bottom of the sea and plucked up the Data Recorder.

A major task had been accomplished. The trick now was to safely transport the two black boxes so that their valuable information could be analysed. Both boxes, returned to shore on 12 July by the *Leon Thevenin*, were placed in water-tight containers and shipped to India where they would be analysed by a study group of experts.

In Bombay, the two devices were kept under armed police guard. In accordance with the orders of Judge Bhupinder Kirpal, appointed by the government of India to launch a judicial inquiry, the boxes were opened on 16 July in the presence of the US National Transportation Safety Board and the Canadian Aviation Safety Board, with Indian appointee Satendra Singh acting as group leader. The experts prepared readouts from the boxes and sound spectrum analysis was carried out at India's top scientific institution, the Bhabha Atomic Research Centre in Bombay.

The experts were in for a surprise. Both the Cockpit Voice Recorder and the Data Recorder had stopped recording simultaneously at 7:14:01. The explosion had severed the power supply of both the devices, which are located near the

103

aft luggage compartment just behind a passenger entry door.

Nevertheless, analysis of the Data Recorder would provide vital clues about the mechanical aspects of the flight and enable investigators to rule out other possible causes of the crash. The data would also confirm that Flight 182 dropped like a stone from the sky. There was no warning. No emergency declared. Nobody had a chance.

The following observations were made by the experts from their analysis of information from the two recorders:
* The aircraft was steadily cruising at latitude 51 5 North and longitude 12 49 West until 7:14 GMT at a ground speed of 519 knots.
* It was on its assigned flight level of 31,000 feet and there was no change in its altitude until the time the data recorders stopped functioning.
* The 747 was flying at a constant heading of 98 degrees, its assigned path towards London which lay to the east.
* The computed air speed varied from 287 knots to 296 knots, sometimes exceeding the recommended air speed restriction with a fifth engine mounted. But Boeing said the slight increase in speed could not have contributed to the accident.
* The plane was in 'clean configuration' – Technical language for flight conditions while a plane is cruising steadily. The landing gears were up, flaps were in retracted position and spoilers used to slow the aircraft were in non-deployed position.
* The data of engine functions indicated that all the Boeing jet's engines were working normally until 7:14 GMT. However, a split-second before the recorders stopped, the No. 3 engine's throttle position was recorded as having moved from 38.65% forward to 43.22% reverse. But split-second recording of other functions of the same engine showed no change. Investigators concluded that the data showing the engine throttle lever in reverse was erroneous.
* The plane was in auto-pilot mode and the data showed no control input from the pilots. There was no movement of the horizontal stabilizer or elevator. The pitch attitude, the angle

of the nose, and the angle of attack of the aircraft did not show any variations. There was no control wheel input by the pilot nor any roll recorded.

* The rudder position was shown as having been about 11 degrees to the right constantly throughout the flight to compensate for the drag caused by the fifth engine mounted on the left wing.

Chief investigator Khola of Air India drew the following conclusions from the analysis of the 'Black Box' data:

'From the correlation of the recordings of the Data Recorder and Cockpit Voice Recorder and the Air Traffic Control (Shannon) tape, it is observed that the beginning of the abnormal sounds recorded on ATC Tape (the split-second burst of microphone clicking picked up by Shannon as the plane disappeared from radar) coincides with the timing when the DFDR and the CVR stopped recording. The correlation further shows that the conversations in the cockpit were normal and there was neither any warning nor any emergency declared till the time the flight recorders stopped functioning.'

The Cockpit Voice Recorder taped the conversation between Captain Bhinder and Flight Purser Dinshaw. Bhinder was asking for that little boy who wanted to see the cockpit. In the final seconds prior to a bang being recorded, Flight Engineer Dara Dumasia was asking Bhinder to contact London Operations for seals to close up the bar aboard the aircraft for customs purposes. He was cut off in mid-sentence by the bang. At the same time, the Shannon ATC tape had recorded indecipherable sounds for a split-second.

Six experts – Dr S. N. Seshadri of the Bombay Atomic Research Centre, Satendra Singh of the Civil Aviation Department in Bombay, John Young and Paul Turner, the two US experts, and P. De Niverville and B. Caiger of the Canadian Aviation Safety Board reached this conclusion about the sounds recorded at Shannon:

'It appears that the ATC recording contains the beginning

of the aircraft breaking until power is lost to the transmitter since channel one and channel four (Captain's and co-pilot's mikes) appear to contain a transmitted signal on the Cockpit Voice Recorder.'

Paul Turner, an expert on Cockpit Voice Recorders, made this individual finding:

'During my observations of numerous Cockpit Voice Recorders I have heard and observed a number of aircraft breakages due to various causes. In this case, the explosive sound on the Cockpit Area Mike occurs prior to any electrical disturbances observable on the selector panel signals. Electrical disturbances can generally be seen prior to audio signal when explosive sounds originate at any significant measurable distance from the microphone (15 feet) and in the area where there are significant electrical systems. It is my opinion that an explosive event occurred close to the cockpit.'

Turner added that 'the cockpit area mike signal that follows the explosive event shows a very much higher noise level than cockpit ambient of 85 decibels, indicating to me that the cockpit area was penetrated and opened to the atmosphere. The selector panel signals show signatures similar to those of an aircraft breaking up and are apparently caused by electrical systems disturbance (circuit breaker blowing, fuse switching etc.).'

Turner concluded: 'The lack of a Mayday call and the apparent inadvertent signal from the cockpit suggest crew incapacitation.'

In a nutshell, the observations pointed to one thing only. When the Cockpit Voice Recorder stopped functioning, an explosion had occurred on board the aircraft while it was cruising steadily at 31,000 feet with all of its functions normal. After that the plane was unable to sustain flight. It lost altitude, broke up in mid-air and scattered bodies, bags and metal. The emergency had developed so fast that the crew had no time even to relay a distress call.

8

Bag Bomb Versus a Jumbo Jet

The Boeing 747 jet airliner is called a Jumbo for good reason. When the Seattle-based Boeing company first put it out for test flights in December 1968, nothing like this mammoth had ever flown in the sky. You get a fairly good idea of its gargantuan size if you stand somewhere on the ground near the nose and look up at the cockpit. The wing-span of the 747 is 195 feet, its height is 63 feet and it measures 231 feet from the nose to the end of the tail section. Depending on the seating configuration, anywhere from 400 to over 500 passengers can be accommodated. Each of its four engines produce a thrust of up to 47,000 pounds to help carry the behemoth's maximum weight of 710,000 pounds into the sky. That's how big this machine of light metal alloy and glass-fibre actually is.

And it is equally complex. The fuselage is so big that it is put together in three separate sections which are then joined together. Other sections such as the wings and the upper flight decks are also attached separately. Before the first test flights occurred, Boeing put its pride and joy through many other rigorous tests.

It was, they said, the toughest airliner ever built. But the larger flying machines get, the lighter their structure. The 747 makes extensive use of titanium and honey-comb panelling of glass-fibre and light aluminium alloy.

Experts also said the 747 was fool-proof. It had back-up

systems for virtually everything that could fail: four engines which operate independently and are fuelled independently, dozens of hydraulic systems and power cables, each with back-up should one system fail.

The 747 is made for flying, to sustain flight at high altitudes, high speeds and greater passenger and payload capacities than ever before. The aircraft, without a doubt, is one of the best ever built by man.

But the Jumbo wasn't made to take on a bomb. The aircraft that's bomb-proof hasn't yet left the ground. That's the verdict of British aviation sabotage expert Eric Newton.

'Designers don't design aircraft to be blown up by a bomb – whatever [their] size,' says Newton, the top international expert in the world today on the effects of explosives on planes. 'Airplanes are made to sustain aerodynamic loads. They are not designed to cope with a bomb, I mean if you were to design an aircraft to be bomb-proof it wouldn't ever leave the ground. They are only made of light alloy and they've *got* to be light.'

Newton should know. Over 36 years, while working for the UK Accident Investigations Branch, he has probed several hundred air accidents, including numerous cases where aircraft either have been destroyed by or have just barely survived explosive sabotage. For his work on aircraft safety Newton was decorated with the Order of the British Empire in 1951 and the companion Imperial Service Order in 1971. Investigators know he's the man to call in when you're probing the suspected sabotage of aircraft. And they'd need his expertise this time too. He would be called out to examine massive amounts of floating debris found off the Kerry Coast shortly after the Air India disaster.

Newton had been almost prophetic in an article which he wrote for the International Journal of Aviation Safety in March, 1985, in which he discussed historical examples of bombings of civilian aircraft over the past 38 years.

'The loss of one large wide-bodied jet transport can represent a financial loss of more than 70 million US dollars,' wrote Newton. 'And an even greater loss morally and commercially to the operating airline, the government and the fare-paying passenger. The loss of one modern aircraft in the future may have far-reaching social and political consequences.'

Therein lay the apparent motive for the terror wreaked on Air India: to teach the perpetrators' enemy, India, a lesson. Deal it a heavy blow by blowing up two of its fleet of ten jumbos. One in London and one in Narita. The motive would have been revenge.

Newton now lives in retirement in Hastings, Sussex. From there he issued his verdict on how well a Jumbo would cope with a bag bomb. And he also shed some light on the question of why aircraft provide a fascinating target for saboteurs. The Boeing 747 is a huge plane, almost the size of a big building, regarded as the toughest flying machine man has ever built. How could a bag bomb down an airliner that's over 240 feet long and 63 feet high? Here is what Newton said:

'If the explosive charge is big enough, let us talk in kilograms of high explosive, say three kilograms of high explosive anywhere in the back of that aircraft [the Air India Boeing] would cause a serious structural failure and it would possibly take the whole back end off.

'If the bomb is in the baggage compartment, which is a pressurized compartment – just over eight pounds per square inch at over 31,000 feet, its highly loaded at that altitude – if you get a massive explosion there, even if it didn't cut the aircraft in half, even if it made a hole in the aircraft of more than about 30 sq. ft., it would obviously be a disaster.

'Now if [the bomb] was located in the front luggage compartment, it's even worse because very near the baggage compartment is the electronics bay,' Newton added.

But wouldn't the metal screen between the forward luggage hold and the electronics bay provide shelter?

'It's very flimsy, it's only a light aluminium structure. It's very flimsy when we are talking about explosives. The only thing that does shield the explosive is the container. All the baggage in the large aircraft goes in containers. Now they're not very strong [either], of course.'

Newton said that a powerful explosive device in the front

baggage compartment would doom the aircraft, without a doubt. It would cause a massive structural failure. And make it impossible for the crew and passengers to survive.

Why would the Cockpit Voice Recorder and the Data Recorder stop monitoring at the same time? What was the indecipherable sound that Air Traffic Control at Shannon picked up during the last moments of Air India Flight 182?

'The sound could be the structural failure of the aircraft,' Newton said. 'Some people would describe it as explosive decompression of the fuselage. It is what I would describe as decompression roar.

'Both recorders are situated right at the back of the aircraft, and the cables run underneath the floor from the electronics bay. Any break-up of the cables anywhere along the floor would cut them off quite quickly.

'If the bomb was of sufficient charge,' Newton added, 'it doesn't matter where the bomb was located. It would be absolutely devastating.'

Why are terrorists so fascinated by aircraft? Why do they enjoy the spectacle of destroying innocent human lives?

'The saboteur thinks his evidence is going to be destroyed, and no one will know what happened to the aircraft. This is the criminal mind thinking that all the evidence that I've put in there will be destroyed. There are many ways of destroying an aircraft and a bag bomb is one of the simplest.

'This is not the first aircraft that's been destroyed by a bomb,' Newton continued. 'There have been two previous incidents of 747s with bombs but they were not of sufficient charge and the planes managed to land.'

Supposing the bomb on the Air India flight was located in the aft compartment, what controls would be left to pilot?

'Not very many,' Newton said. 'The main controls run through the roof of this aircraft. The tail-end controls run through the roof, but there comes a time when they have to come downwards. You'd only have to open up the shell of the aircraft and the plane would start a dive and go out of control.'

Would an extra engine mounted on the port wing contribute to the loss of control?

'I don't think so,' was Newton's opinion. 'I don't think this

110

has much to do with it.'

Newton said he had listened to the ATC tape in Shannon and felt it contained garbled human voices along with a huge roar from the aircraft as though air was hissing out through an opening in the fuselage. 'It sounded like the air was rushing out because it [the fuselage] had been ruptured,' he said.

'I can tell you quite clearly that there was no explosion on the flight deck,' he added. 'I've seen the flight deck door and some of the structures. Absolutely clean as a whistle – the door was still in a locked condition.'

From the pattern of wreckage he had seen, where would he say the bomb was located?

'Well, from my experience, it would be located most likely in some baggage compartment. I can't say whether it was in the front or the back.'

In Vancouver, the Mounties were now getting criminal evidence which indicated that whoever planted the bomb on CP Flight 003 going to Tokyo, also planted the Air India bomb. The inference could, therefore, be drawn that both devices would carry the same amount of explosive charge. Narita police had made a series of tests, and told the Mounties that while the bomb would not be so devastating on the ground, it was sufficiently powerful to have downed the CP Air 747 had it exploded prematurely while the plane was in flight, regardless of the location of the bomb on the aircraft. Yoshiaki Saito, a top member of the Japanese police team probing the Narita blast, also told them that the bomb would have proved fatal for the CP aircraft had the plane been delayed on its trans-Pacific flight with L. Singh's bag bomb aboard. The 390 passengers and crew of the Canadian Pacific Flight 003 had made it to Tokyo on a wing and a prayer, quite literally.

If the Mounties wanted historical evidence of the devastation bombs have caused aboard civilian aircraft, Newton could provide them with 58 known examples of explosive sabotage of airlines in 38 years.

'The detonation of an explosive device, such as a bomb,

within civil transport aircraft is fortunately a rare but often a disastrous occurrence,' Newton wrote in the Journal of Aviation Safety. 'Unfortunately, world-wide, over the past 38 years, some 863 passengers and crew [have] lost their lives [owing to explosives] and 58 aircraft were damaged or totally destroyed [hijacking excluded].'

Of the 58 aircraft which were hit, 24 were completely destroyed. Eighteen aircraft were of the piston-engine type while 40 were modern turbine-engined types. Newton could now add the Air India toll to his count of 863 fatalities in 38 years. The Air India disaster and the Narita blast would add a total of 331 to his list, Air India the single worst case of sabotage of a civilian aircraft ever known to history.

Newton said in his article that the explosive disasters occurred in 30 countries and affected 31 different airlines. In only six of these 58 cases he had looked at had he seen any evidence of fire as a result of an explosion. The invariable cause of disaster had been the tearing of a hole through the aircraft's thin-skinned fuselage.

But the Mounties like to see for themselves. At a secret location in Canada, explosives experts experimentally blew up the metal shell of the container similar to what the terrorists had used to carry their deadly device to Narita. It was the aluminium shell of a stereo. They blew up a whole series of similar stereo containers, trying to determine how much explosive would be needed to create the same size particles of debris that Japanese police had shown them.

They progressed from one stick of dynamite to eight and then more. By re-creating the Narita explosion as exactly as possible, they were able to figure out the explosive power of the Narita bomb. Then they wanted to see how much damage an explosive of that strength would do to a luggage container similar to ones used on Air India Flight 182. They tried airline baggage containers with nothing in them. Then they packed them with bomb-rigged bags and blew them off. They put the bag in the centre of the container surrounded by other bags to sandwich the explosion. They tried all sorts of dynamite, again ranging from one stick to several.

It worked every time with just a few sticks of dynamite. They tried it dozens of times. Every time, the blast left a gaping hole in the container. They were surprised by how little explosive power it took to make a hole in a luggage container and the fuselage of an aircraft.

A hole in the fuselage is just what an aircraft doesn't want, because the cabin and the luggage compartment are like a pressurized aerosol canister. When the aircraft is flying at 31,000 feet, the cabin pressure is equal to normal atmospheric pressure at 10,000 feet or less.

The Mounties had seen for themselves the devastation a bag bomb can cause. There was no contest. The bag bomb won hands down against conditions identical to those found on a 747.

Newton was right, no plane has left the ground that's bomb-proof.

9

List of Suspects

The Mounties had an early lead from the Canadian Security Intelligence Service which pointed out a group of militant Sikhs as possible suspects. Crime laboratory investigations connected them with explosives just days after the Air India crash and the Narita explosion. But that information, while a vital lead, was definitely not the complete answer.

Traditionally a criminal investigation calls for the drawing up of all possible scenarios. No matter how wild. Then you weigh the possibilities and probabilities. Who would want to blow up two Air India jets? Clearly that was the question that needed to be answered, because the Mounties knew that in both cases where the Singhs had not checked in for their flights but had got their baggagge aboard, they were connecting with Air India. M. Singh, who was to leave from Vancouver, was connecting with Air India 182 in Toronto. L. Singh was to connect with Air India Flight 301 for a flight to Bangkok from Tokyo.

However, although the finger pointed at terrorism, the pattern didn't fit. For example, in the Middle East, in any given incident involving the Palestine Liberation Organization or its off-shoots, or for that matter, some of the European terrorist organizations, the terrorists have always claimed responsibility for actions such as hijackings or sabotage. That is the whole point of terrorism: Punish the enemy and then tell the world, 'We did it!'

But in the case of Air India and the Narita blast, no one had come forward with a serious claim of responsibility. Yes, there was one solitary claim in an anonymous call to the New York Times in which the caller said the bombings were the work of the International Sikh Students Federation. But that claim was dismissed by intelligence agencies in both the United States and Canada as a prank.

The Mounties then drew up a chart showing who had the most to gain by blowing up the Air India jet. Certainly, Sikh radicals would have achieved a big score if they had succeeded in blowing up two Air India Jumbos almost simultaneously in Narita and London. For months, in Sikh temples in London, Vancouver, Toronto and other cities advice was being passed around that those who believed in the cause should boycott Air India because it was a symbol of the government of India, which they said was hell-bent on destroying the Sikh religion. Two out of the fleet of ten jumbos owned by Air India would have been a feather in their cap.

Initially, though, the vast majority of Sikhs in Canada who do not endorse violence thought that most of the victims were Sikhs. At first the RCMP also believed that. Why would Sikhs blow up fellow Sikhs in the air? Wouldn't that only serve to create a backlash against militants from the massive majority of moderates in Canada and elsewhere? Furthermore, when the reservations were made, telephone numbers of Sikhs and the main Sikh Temple on Ross Street in Vancouver had been given as a supposed means of contacting the customers.

The Ross Street temple, which boasts the highest membership of Sikhs anywhere in the North American continent, supports the cause of a separate republic for the Sikhs. But it does not endorse violence as a means of achieving that aim. The temple was stunned by the loss of life and immediately condemned whoever was responsible for the disaster. In addition, the temple executive declared 15 days of prayer for the victims. Why would Sikh radicals malign the temple by leaving its telephone number for investigators to connect with acts of terrorism?

Ever since the tragedies, and long before that, a few

members of the Sikh community have alleged that Indian intelligence agents have been operating in several Canadian cities with large Sikh populations. Everywhere the Canadian Security Intelligence Service went to interview known Sikh radicals, they were directed to look at RAW – the Research and Analysis Wing, the external arm of the Indian spy agency. So CSIS passed on the scenario Sikh radicals were painting to the RCMP.

Let us suppose for a moment that it was elements of RAW who blew up the Air India jet. What would they have achieved? For years, India has been pointing at Canadian, British and American Sikh elements as the breeders of turmoil in India's troubled Punjab State. And that is partly true. But India has also maintained that authorities in Western countries were doing nothing about it. Doing nothing to stem the flow of money, and even in some cases weapons, to the extremists in the Punjab.

For example, the Indians have been pressuring Prime Minister Margaret Thatcher to put her foot down on Dr Jagjit Singh Chauhan, the self-proclaimed President of the Republic of Khalistan, who has been doing what he wished under the shelter of democracy in Britain. India has also been turning the diplomatic screws on Canada and asking that it send Vancouver fundamentalist Talwinder Singh Parmar, wanted for the alleged murder of two police officers, back to India to stand trial. (The murders allegedly occurred in the Punjab village of Dehru in 1982. The Indians were unable to prove their charges when Parmar was detained in Germany on an Interpol warrant. He was set free in 1984 after being detained for one year.) India also said Parmar and his top aides, like Surjan Singh Gill in Vancouver, who had earlier aligned himself with England-based rebel Chauhan, were able to get away with inflammatory speeches against India and open support for the secession of the Punjab from India.

Well, Thatcher was listening, especially when Chauhan made suggestions that suicide squads would be formed for missions inside India. His remarks came after he pointed out, in the wake of the assassination of Prime Minister Indira Gandhi, that traditionally anyone who desecrated Sikh holy shrines as she had done did not live very long. The innuendo was that Rajiv, her heir, wouldn't live long either because he represented the same dynasty.

In the United States, the FBI, in an operation code-named Rite-Cross, uncovered a plot to assassinate Rajiv Gandhi during a trip to Washington in the summer of 1985. At that point, the extent of the radical Sikh network became apparent to the Reagan administration. The Secret Service and the FBI had begun taking a look at various Sikh safehouses in California, Chicago and even Washington. The foiled murder plot by fugitives Lal Singh and Ammand Singh had achieved for India the thing it wanted most: surveillance of top Sikh militants in the United States.

Canada, though, according to India, was still not listening. So what if RAW or rogue elements within the organization planted the bomb on Air India to drive the point home? That would mean Canada finally would have to act, Sikh radicals argued. That it could no longer allow them to operate freely in this country.

It was a possible scenario, although not a probable one. It had to be explored, nevertheless. There were at least three diplomatic bags on the downed Air India plane that had not been checked because of diplomatic protocol. Air India passenger agent Yodh, based in New York, had made his first trip to Toronto to load this particular flight. The wife of Toronto Indian consul Surinder Malik had cancelled her flight. There were rumours in Vancouver that the wife of Vancouver Consul Jagdish Sharma had cancelled her flight too. Furthermore, because of the devastation the incident had caused, it appeared to be a polished, professional job. Where would militant Sikhs get the technology to make such a sophisticated bomb? Such was the basis for the Indians-did-it-themselves scenario put forward by the militants.

The Mounties questioned the Vancouver consulate security man who had packed the diplomatic bags, which weighed just over 10 kilograms. They questioned some Air India employees about their knowledge of explosives. Indian diplomats were astounded and offended. But they understood that the questions were necessary in a criminal probe which hinges on the elimination of possible causes until only one, the true one, remains.

The diplomatic bag theory was discarded almost immediately because the bags had been flown from Vancouver two days prior to being put aboard the Air India jet by Air Canada. Furthermore, they had been stored in Toronto before being loaded onto the plane. If they blew up prematurely, what face would India show to the world? And if the Indians had decided on blowing up one of their own planes, why would they use the diplomatic bag route? Mrs Malik had booked several flights to India. She had not made up her mind to travel yet. It was an either–or situation.

I asked Sharma in Vancouver about his own travels.

'My father died and I went to India on 18 May,' he explained patiently. 'My wife came afterwards, on 8 June.'

'Let me tell you,' he added. 'If Indian Intelligence *had* made such a horrible decision – to sacrifice 329 lives – they would have said to hell with a diplomat's wife.' And furthermore, Sharma said, 'Given the propensity for Indians to talk, you tell one, you've told 20. The basic tenet of an intelligence operation is secrecy, utmost secrecy.'

The RCMP queried Yodh about the reason for his presence in Toronto for the first time and found that he was replacing the regular Air India representative Sarwal Ashwani, who was on vacation.

Furthermore, a top government official from India, no less than a state education minister, had flown to Montreal on the same flight. If the wives of Indian diplomats had been warned, wouldn't he have been warned too? If Yodh or D'Souza had been aware of the bomb, they would never have dared to fly to Montreal with AI 182 from Toronto as they did.

The Indians-did-it-themselves theory was dismissed almost as soon as it was entertained. The criminal investigation in Vancouver was pointing in an entirely different direction. And besides that, the Canadian Security Intelligence Service had no evidence to indicate that any of the suspects the Mounties now had were in any way, shape or form connected with the Indians. On the contrary, the early leads pointed to men who hated India with a passion. They were fundamentalist Sikhs who were hell-bent on revenge and who had openly vowed to get even with India for the assault on the Golden Temple, and for the New Delhi massacre of Sikhs in the wake of the assassination of Gandhi.

The Mounties took another look at the Air India passenger list and found less than 50 of the passengers on the flight were Sikhs. They also learned that some militants had been branding those fellow Sikhs travelling by Air India as 'gadars', meaning traitors. Several had even been threatened for travelling with the airline. And the word was out that 'if you know what's good for you, you won't travel with Air India.'

The aim of the terrorists had also become quite apparent from a study of the timetables of the flights. Evidently nobody had claimed responsibility because the operation had been screwed up. It became apparent that the terrorists were actually trying to blow up two Air India jets *on the ground*. One in London and one in Tokyo. There was no doubt that the explosions would have caused some loss of life if everything had gone according to plan. But it didn't. Instead of getting just the planes, and probably a few unlucky airport workers, the perpetrators had managed to blow away 331 people. Who wants to be labelled with that kind of a horror? That's why there was no credit taken for the sabotage.

The timetables showed their plans precisely. The Air India timetable, easily accessible to anyone, shows that Flight 182 was scheduled to touch down at London's Heathrow Airport at 7:45 in the morning on Sunday. But it had been delayed by over an hour in Toronto. The explosion happened aboard the aircraft at 8:14 London time. Had everything gone according to plan, and the plane had not been delayed, the aircraft would have blown up almost 30 minutes after arrival at Heathrow. Most, if not all, of the passengers would have disembarked and so would the crew.

The same time factor was evident in the case of the Narita explosion, exactly one hour before the Air India blast. CP Flight 003 was scheduled to arrive at Narita at 2.55 p.m., although it arrived slightly early. However, the bomb went off just after 3:20. Again, there was a time difference of between 25 to 30 minutes after the scheduled time of arrival. The RCMP believe that the terrorists were hoping that because the Air India Flight 301, for which the bomb was meant, was taking off two hours after 003 arrived from Vancouver, the bomb would make its way to the plane within 30 minutes.

It is also possible that the terrorists miscalculated – and that wasn't the only mistake they'd made – on the timing of the Narita blast. They may have timed the bomb to blow up one hour too early. Had they given it one hour more, they would have achieved simultaneous explosions in London and Narita. They'd have 'taught the enemy a lesson'.

The time factor also blew a hole in theories by pundits that the Air India bomb had to be carefully placed to blow the plane up the way it did. The plane was supposed to explode on the ground, so it didn't matter where they placed the bomb. The correct placement theory was academic.

The Mounties began concentrating their efforts on radical Punjab separatists. They hit pay-dirt in no time. There was a trail of evidence. It really *wasn't* a professional job. It was the work of people who were taking on an operation of a size and complexity which they had never undertaken before. And they hadn't attended schools of terrorism to learn how to wipe out the evidence afterwards.

The perfect crime hasn't been committed yet, police say. And this operation, too, was considerably less than perfect.

10

Phoney Phone Trail

One of the most tenuous leads the RCMP were left chasing in Vancouver was the telephone number on the ticket files of L. Singh and M. Singh at the CP Air reservations office in Vancouver. The man who had booked flights with Martine Donahue at CP Air, initially for Mohinderbell Singh and Jaswand Singh on 19 June, had left one phone number – and although the names and initials and details of flights had changed, that number had remained constant. It was 437-3216, and it was to lead the Mounties to some surprising results.

The task of tracing the telephone number was handed to Constable Sandy Sandhu, a Vancouver Sikh himself, who was to be an invaluable member of the huge RCMP Task Force probing the two disasters. A quick check with British Columbia Telephone Company led him and fellow officers from RCMP headquarters to the door of Hardial Singh Johal in East Vancouver.

When the constable knocked on the door of the new East-End home, Johal seemed surprised to have a Mountie on his doorstep. Sandhu asked if he could come in for a brief chat. Johal thought it over quickly. He had heard that the RCMP were scouring the Lower Mainland, interviewing dozens, perhaps hundreds of Sikhs in the Vancouver area looking for clues to the twin disasters. It was the first time in his twelve years in Canada that a policeman had appeared at his door.

But the man, known as a moderate inside the Sikh community, was not about to shut his door on the Mounties. He ushered Sandhu inside his home and led him upstairs to the living room.

Although this was his first brush with the law in Canada, the Sikh activist had been in jail in India three times. He'd spent his first night behind bars in 1960 when he was barely 14 years old. That time he'd paid with seven days of his life for his beliefs. Then, Sikhs had been demanding changes in the way the government in New Delhi had been appointing representatives to the temples throughout the country. In 1971, the struggle against government appointees was still going on, and Johal landed in jail a second time while working as an engineer with the supreme authority in Amritsar that manages Sikh temples, including the Golden Temple. This stay in jail in support of the cause and his faith would also be brief.*

In 1972, Johal moved to Canada. The bright, energetic engineer was looking forward to a new career and a whole new life in his new land. But he had not forgotten India or the Punjab. He returned to his homeland in 1982 to show his solidarity with Sikh leader Harchand Singh Longowal, the moderate leader of the popular Shromani Akali Dal party of Punjab. Longowal had even stayed in Johal's house in Vancouver during a two-week visit in 1979.

In August 1982, as Johal flew to the Punjab, Longowal was in the middle of his *Dharam Yudh* – religious war – against the Indira Gandhi government. The so-called holy war, one in a series of agitations organized by Longowal, was designed to force the government of Mrs Gandhi to bow to a package of 45 Sikh demands. Some were of a religious nature, others involved the better distribution of water for the crops, maintaining the former quotas of Sikhs in the military, and the extension of Punjabi as a second official language in neighbouring states.

The 40-year-old Vancouver school board engineer gladly courted arrest as part of the campaign to clog India's jails and court system, and along with thousands of other Sikhs, spent a few nights in prison. At about the same time that Longowal launched his agitation, two potentially explosive hijackings rocked the country within a year. First a Sikh extremist hijacked a domestic Indian Airlines flight and attempted to land in Pakistan. But India's moslem neighbour to the west refused to allow the jet to land. The hijacker forced the plane back to Amritsar in the Punjab, where he was eventually arrested and his 126 hostages set free. Two weeks later, another Sikh hijacked another Indian Airlines jet. Once again, he sought refuge in Pakistan. Once again, Pakistan refused him succour. And once again the incident ended in Amritsar, the Sikh holy city. But this time, the hijacker was shot and killed by police.

Such was the atmosphere when Johal joined Longowal's *Dharam Yudh*. Comradeship in adversity formed strong ties, and Johal, the man of two countries, became very close to Longowal. But in a few years, divisions inside the Sikh religio-political movement – manifested in sharp splits between moderate-militants and extremists – had begun to show. While Johal was in the Punjab in February 1985, mourning the death of his father, he was honoured by the presence of the head priest of the Golden Temple, Giani Kirpal Singh, as one of those who had come to show his grief. But shortly after the priest left Johal's house, Giani Kirpal Singh was gunned down by unknown assailants. Fortunately, the priest survived the attempt on his life.

Longowal, however, had not been so lucky. In June 1984, he had surrendered to the Indian Army just before the Golden Temple was stormed to flush out armed followers of Sant Jarnail Singh Bhindranwale. The surrender itself was viewed by extremists as a sell-out to the government. They argued that if Longowal was a true Sikh, he would have stayed inside and been martyred. But Longowal had now begun to see that the agitation he had begun had snowballed into violence and anarchy. So Longowal made his peace with Rajiv Gandhi and signed an accord. It cost him his life on 20 August 1985, when he was gunned down inside a temple.

During the last few months prior to the Air India disaster, Johal had sided with Longowal's peace initiative, a move which of course didn't earn him any points within powerful militant groups in Vancouver.

For that support, he had already been branded either as an Indian agent or a make-believe Sikh. He had enough trouble already. And now what did this RCMP officer want from him?

Plenty. Settling down in Johal's home, the Mountie pulled out a sheet of paper with several telephone numbers jotted down on it. Johal glanced at them. Then the policeman asked a question which almost floored him.

'Do you have any idea why your telephone number was used by L. Singh and M. Singh when their tickets were reserved? Do you know them?'

The number didn't lie, and Sandhu showed it to Johal. It was 437-3216, his former number – the one he'd left behind when he had left his old residence ten months earlier. He stared at it.

'That was my number ten months ago,' agreed Johal. 'I don't see why anybody would use my number.'

Sandhu explained that it was the number the two shady passengers had left with CP Air on 19 June, just four days before the Air India disaster and the Narita explosion. Despite all the changes in flights and schedules and bookings, the terrorists had stuck with that number. Astonishingly, the last four digits were the same as Johal's present number, too.

Johal agreed that there were three possible conclusions the RCMP could draw from this piece of information. The first was the obvious; that it was he who had booked the flights connecting with Air India. That he had, when asked point blank what his number was, thought of his former number and left it with the reservations agent. The second scenario was a fantastic coincidence: someone had made up a number and by chance it happened to be his. The third was less coincidental and more probable – someone had tried to frame him because of his support for the peace accord.

Those were the possibilities. The first two scenarios didn't stand up very well, but the third was worth considering.

Johal defended himself. His number was a very public one, he said. After all, he was very active in the Sikh community in Vancouver. He'd joined a 'back to basics' movement in 1979 as part of a campaign to wrest control of the main temple from men he felt did not adhere to basic doctrines of the Sikh faith, namely keeping unshorn hair and beard, among other things, and coming into the temple without their heads covered. From 1980 to 1982, he'd served as the temple treasurer, and his home number would have appeared on most letters leaving the temple. If he had been involved with such a plot as the Air India bombing, he would hardly leave a number which could be traced back to him, he told Sandhu.

'Every criminal tries to protect himself,' said Johal. 'My number was used by an enemy, not a friend.'

Johal had stubbornly stuck to his moderate beliefs, and he certainly had made enemies. He even fought the pro-separatist Sikh executive of the Vancouver temple in a bitter election contest, although he finally bowed out. Some of Johal's enemies attacked him a few months later.

Leaving the Vancouver school where he works, carrying a ladder to hoist the flag, Johal was accosted by three men brandishing iron bars. They greeted him in Punjabi and said they had a present for him – a little something for supporting the peace accord in India. The first blow glanced off his turbaned head. The second bruised his foot. Johal then managed to run indoors and wrest a fire extinguisher off the wall. He turned it on his attackers while flipping the fire alarm switch. His assailants fled before serious harm came to him.

The Task Force had run into a dead-end of sorts. Johal was a self-styled moderate. Why *would* he leave behind a number that could easily be traced back to him? It didn't make sense. So the police turned to the possible permutations of the telephone number. This raised another incredible and intriguing possibility. If you change the number's last digit, you get 437-3215. And that is the present number of Sikh Sodhi Singh Sodhi. He became the next man to whom the Mounties would pay a visit.

Sodhi had already turned up in police books once. On 4 June, 1984, Sodhi and his close friend Jasbir Singh Sandhu, now a regional representative of the London-based Khalistan National Council, stormed the Indian consulate in Van-couver. The two men burst into the Howe Street office screaming slogans and brandishing a sword. Panic-stricken employees were ducking for cover and climbing out of windows. But Sodhi wasn't out to hurt anybody. He just wanted to tell Mrs Gandhi that Sikhs world-wide were appalled at her decision to raid the holiest of their shrines, the Golden Temple.

Sodhi and his partner grabbed a portrait of Prime Minister Indira Gandhi and smashed it on the ground. Then they laid a pair of shoes on the tattered photograph – a gesture considered to be a supreme insult in India.

Sodhi defended his actions later by claiming they were a spontaneous reaction to the storming of the Golden Temple in Amritsar. It had happened in the heat of the moment. He insisted he meant no harm.

'I heard on the radio that the army had moved into the Golden Temple and that about 400 Sikhs had been killed,' said Sodhi. 'We didn't want the military in the Golden Temple. If they had problems they could solve them through negotiations. We knocked on the door of the consulate and they opened it. We told everybody to go inside their offices and asked for the consul.

'We took a picture of Indira Gandhi and put shoes on her picture. She's the one who insulted the Golden Temple. So we insulted her.

'The man who opened the door was trying to jump out of the window,' he said. 'Everybody was scared because we had a sword. But we didn't use it. We were just yelling, just slogans.'

The pair surrendered to police without a struggle, were handcuffed and driven away in a paddy wagon. The Vancouver Indian Consul declined to press charges against them, hoping to cool the passions of Sikhs in Vancouver.

Now, on the first of many visits the RCMP would make to his house, Sodhi too was being asked if he knew L. Singh and M. Singh. The Mounties were there on the basis that if a person gives a false number, he's likely to leave a number that is at least traceable to himself.

Sodhi, however, was completely perplexed by the questions he was being asked. He offered to take a lie-detector test and said he would never endorse a barbaric operation which cost the lives of 331 people. But the RCMP came back to him repeatedly, saying he also looked like the person who had picked up the tickets from CP Air.

'I offered to go in front of the agent,' Sodhi said. 'A policeman called me later to ask if I would be willing to take a lie-detector test. I replied, "You can come and take me any day you like. I have nothing to hide."'

Sodhi was right. The phone number mystery was still unsolved. But there was one more number left to explore. The last number left on the RCMP's list was a South Vancouver number, 324-7525. It was the number left by whoever booked A. Singh on CP Flight 003 from Vancouver to Tokyo, with a connection to Bangkok via Air India for 22 June. Although A. Singh had held a confirmed booking for the flights, he'd never purchased a ticket.

The number wasn't very hard to trace. It turned out to be the main office of the Ross Street Sikh Temple. When questioned, the temple executive denied any knowledge of an A. Singh. The telephone number appears on every letter sent from the temple and is listed in the phone book. Anybody could have used it. The number trail had gone cold.

There were two possibilities; two theories to explain the riddle of the ticket telephone numbers. Could Indian agents have left them as part of a bid to discredit the Sikhs? But this first possibility had been thrown out by the Mounties because of convincing criminal evidence they were turning up about the involvement of a small cell of a fundamentalist Sikh group. They had become the prime suspects, and it was thought that leaving the phone numbers might have been their way of indirectly hinting at a responsibility they would not have dared to claim openly – even if their operation had gone according to plan and both planes had blown up on the ground.

A group of terrorists planning such a blow to Air India, especially one operating in Canada, would never dare claim responsibility. Canada wasn't Tripoli. The whole nation would be outraged and the full force of the law would be brought down on them. The terrorists were counting on the fact that massive explosions on the ground on Air India 182 at Heathrow and aboard Air India 301 in Narita would destroy clues which could be traced to them. But by leaving a few numbers of fellow Sikhs and of their institutions, they could at least put the credit where it was due. And if in the process, Hardial Singh Johal took the punishment, so be it – from their viewpoint he was a traitor to the cause.

Another scenario put forward by police was that when the first booking was made on 16 June in the name of A. Singh, the caller – not expecting to be asked for a phone number – had left behind the most familiar number, that of the temple.

Afterwards the group had given thought to the aspect of having to leave behind a contact number. The theory goes that they may have felt leaving the number of the temple was a mistake. Therefore they had Johal's former telephone number ready when they made subsequent bookings for Mohinderbell Singh for Tokyo and Jaswand Singh for Air India 182. Both names were changed for some reason but the number had remained the same. Under this theory, too, Johal's number would clearly be a plant and it would explain why a ticket was never purchased under A. Singh's name.

The phoney telephone trail had indirectly provided another clue – clear proof that L. and M. Singh were not genuine travellers, but fronts put up by the terrorists. A bona fide passenger wouldn't give numbers where he couldn't be reached. It was another fragment of the puzzle, and the emerging picture looked more and more like a wheel of terror with its hub in Vancouver.

11

Killer Stereo

It was clear that the inexperienced terrorists had assumed that their bombs would leave little evidence behind, and in fact they were partly right. But three things worked against them: Bad timing, the wizardry of Japanese forensic experts and the willingness of the Mounties to search for a needle in a haystack.

Shortly after the bomb-blast ripped through Narita's modern airport, shaking the terminal building, killing two baggage handlers, injuring four others and scaring the daylights out of passengers, forensic experts started the painstaking task of picking up every minute particle of what was left of the powerful explosive device. Then they began piecing together what they could.

The terrorists hadn't heard about Japanese forensic technology. And they had even helped the Japanese a bit by miscalculating their time. All the evidence pointed to an attempt at simultaneous explosions, one aboard Air India 182 at Heathrow in London and the other aboard Air India 301 at Narita. The terrorists miscalculated, however, by exactly an hour. They would have achieved their aim if they had delayed the timer by one hour on their Narita bomb. But the miscalculation meant that the bomb exploded within the terminal, rather than after the bag was put aboard the waiting Air India jet that was to take off two hours after CP 003 arrived with its deadly cargo. The blast occurred just as the luggage handlers were taking the bag out of a container. Somehow the terrorists had lost an hour in trying to calculate the difference between Tokyo time and Vancouver time before setting their deadly device.

Had the bomb gone off on the Air India plane while it was parked on the tarmac, much of the forensic evidence would have been lost in a ball of fire in the confined space of a luggage container. And this evidently was the plan. Cause a thundering explosion and then start a fire. The device was designed to do exactly that. But because of the error in setting its timer, it had detonated on the concrete floor of the terminal building, with vast space for particles to fly out in all directions. Particles were embedded several feet away in walls. With grim justice, the bodies of the two unfortunate victims of the explosion would also give up vital clues to their murder in the form of tiny fragments of metal, particles of a circuit board and other fragments. Eyewitnesses reported a ball of flame and acrid smoke.

Forensic experts who set to work on the bodies of the two victims, plucking out the tiny fragments, put together a startlingly clear picture of the bomb. It had been placed in a grey vinyl suitcase, and the recovered bits of metal revealed that the explosive had been concealed within the aluminium shell of an am/fm stereo tuner. The experts could even read some of the numbers on the shell. And then, comparing the numbers with various makes of stereos produced in the Orient, they quickly learned that the metal casing was that of a Sanyo tuner. The terrorists, in another oversight that indicated their inexperience, had even placed the tuner in the original box that it had come packed in. The experts then found traces of a particular colour of adhesive tape that had been used to fasten the various components of the bomb – the explosive, the detonator, the timer device and even a lantern type battery.

The news relayed to the RCMP about the brand of stereo used was like music to their ears. It was a major breakthrough. The scenario was easy to follow. The insides of the tuner casing had been removed, to make room for about eight sticks of dynamite and other components necessary to make the infernal device. Finding the fragments of the cardboard container of the stereo was definitely an important bonus, for this meant that somebody probably had bought it new.

Luck was still at work, for the Japanese determined that they were looking at the shell not of just any Sanyo tuner but of a model FMT 611K. The forensic team next wrote to the Sanyo factory about this particular type of Sanyo tuner. Several days later, the factory came back with a surprising reply: All 2,000 of these units made since they were first produced around 1979 had been shipped to Sanyo's warehouse in Richmond, near Vancouver.

How do you track down just *one* of 2,000 tuners of a type that has been discontinued for three years? Do you even try? The question was answered as quickly as it was raised at RCMP headquarters in Vancouver, where Superintendent Holmes and Inspector Hoadley were poring over the task of tracing these 2,000 tuners which had been distributed across the length and breadth of the province of British Columbia. Difficult for sure, but not impossible, they decided. Long shots often hit pay-dirt.

And so it was that Inspector John Hoadley set out from RCMP headquarters for the drive to Richmond, where he was welcomed with open arms by his childhood buddy Cis Oliver, who happened to be plant manager at the Sanyo warehouse. Oliver was delighted to see the man he believed to be one of the finest police officers in Canada. Oliver knew from his school days that Hoadley was not a man who would give up when he faced difficult odds. But 2,000 tuners! Good luck, Oliver said. He knew it would be a tough job, even for Inspector Hoadley.

Hoadley sat down to pore over sales records. He wasn't looking for warranty records, that was for sure. Terrorists wouldn't need to file a warranty card, for they wouldn't need to repair their tuner. For six hours he pored over what records were still held in the Richmond office. But much of the documentation was now in Sanyo's Toronto headquarters, because the stereos had been out of production since 1982. Hoadley promptly phoned and asked Toronto Mounties to start digging at the offices of Sanyo in that city. For two days, his colleagues checked over company records, making photocopies whenever they found sales of model FMT 611K to various discount electronics stores. The haystack was being narrowed, although the needle hadn't grown in size.

Men were dispatched to check each store that had purchased the tuners. Most had run out of stock at least two years earlier. What the Mounties were looking for was a dealer who might have had a few still lying around unsold until recently. The stereo tuner had already become a dinosaur. What they needed was the right stereo unit and the right buyer. One who didn't care about the latest advances in stereo technology, perhaps because he wasn't planning to use the tuner for listening to music.

It was becoming a discouraging task. 'None in stock' was the standard reply. Then, just as they were beginning to feel the frustration of long hours of work and no results, the investigators got lucky. It happened in the sleepy little city of Duncan on Vancouver Island, a ferry-ride and short drive away from Vancouver. A Mountie had just plodded into the F. W. Woolworth store, expecting the standard reply. But what he heard was music to his ears.

The sales assistant told him, 'As a matter of fact, we just sold our last one a few weeks ago.'

Bingo! The Mountie knew he had stuck gold. The long shot had paid off. Once again the terrorists had failed to cover their tracks. The clerk remembered the sale because the store was glad to finally get rid of an out-of-date stereo that had been lying on the shelf gathering dust. And, of course, she remembered the two turbaned and bearded men who had walked into the store to buy it. One of the faces the clerk described was already all too familiar to the Mounties. They'd heard about him from Day One of their investigation.

Why had the men chosen *this* stereo tuner? No doubt for its size. The size was just right for carrying a big load. It measured 16 inches by 14 inches by more than 5 inches. Larger than recent models for certain. The sales records showed that it had been purchased on 5 June, just 17 days before its shell was sent from Vancouver to Tokyo aboard Flight 003 with a bomb concealed inside.

This discovery raised the interesting and very likely possibility that the Air India flight had carried a bomb not only of the same explosive magnitude but possibly inside a stereo unit too. Duncan was combed from one end of the town to the other, but no one else remembered selling a stereo tuner in recent days to two men in turbans. That didn't destroy the theory, however. A second tuner could have been purchased elsewhere or a second-hand unit lying idle at home might have been used. Maybe that was where the idea of putting the bomb inside a stereo came from in the first place.

12

Court of Horror

It was 17 September 1985. The stage was a courthouse in Cork. And unfolding before Coroner Cornelius Riordan was a drama of gruesome, gory horror. Most of the listeners sat in stunned silence. But the coroner had a job to do: to determine just what had happened to all the victims of the Air India 182 disaster who had been brought lifeless to the shores of Ireland. His job was to establish where, when and how they died as accurately as possible; to make some sense out of this tragedy.

'It is my very sad duty today to open an inquest into the deaths of the victims of the Air India disaster,' he began. 'You are all aware of the grim tragedy which overtook the flight on the morning of 23rd June. As it left Montreal for London Heathrow on what turned out to be its last and fateful journey it had, I understand, 307 passengers and 22 crew members. Early on that June Sunday morning it encountered a fatal catastrophe over the Atlantic Ocean off the south coast of Ireland.'

Riordan thus set the stage for the coroner's jury to listen to heart-breaking stories from pathologists and the few witnesses he was able to summon. A lawyer representing the Canadian government tried to impose limitations on how far he could go with his inquest. Ivan Whitehall was trying to draw a line for Riordan. He was, after all, a lawyer who was looking out for the interests of the Canadian government.

That a bomb had been allowed to slip into an airliner that had already, several months before, asked for that government's protection was the last thing Ottawa wanted the world to hear.

A parade of pathologists would willingly cooperate with Riordan, though. And despite the difficulty of getting other witnesses, Riordan would shed some light on what had happened. His first witness was Air India passenger agent Divyang Yodh, who was the man in charge when Air India Flight 182 left Toronto. He had travelled with it to Montreal. He had seen the passengers and crew off on their date with doom. But all Yodh could tell the court was the seating arrangements in the aircraft and the number of passengers in each of its sections. Then it was the turn of Thomas Lane, the air-traffic controller who was monitoring the radar scope when Kanishka vanished from his screen.

'I was at Radar Control at Shannon when I took up position at the radar screen. We had three aircraft in position [on the screen],' he said, describing the events in the control room at Shannon. 'There was between six and 20 miles between the first and last. I was in contact with the Jumbo jet and it was one of three aircraft approaching Shannon Control Zone about the same time. The three aircraft identified themselves and were given coded identifications by Shannon Control. These would enable Shannon Control to lock onto them, using computers in all subsequent transmissions. Air India was the first. There is a transponder code and we gave them their transponder code, which they acknowledged and were immediately identified by the radar controller. He would punch his number in and we would receive that then on our radar scope. That would then record what the altitude readout was and where the aircraft was.'

Lane testified that once the aircraft's transponder tunes in to the Shannon Control radar, the aircraft's position would show up as a diamond with five dots. 'At 8:13 local time, I observed that the AI-182 signal deleted itself from the screen. It did not respond. He was not answering then. The three planes were on the screen for approximately eight to ten minutes when the Air India jet disappeared,' said Lane. 'The position was now serious and they did not reply and I called the aircraft that were behind and that was the TWA and Empress Flight.'

Then when all efforts to reach the Air India plane had failed, Lane said, he realized something deadly had happened. He declared a full-scale emergency.

Desmond Eglington, chief controller at Shannon, testified that the last contact with the aircraft was at approximately 8:09:58 local time (7:09:58 GMT). Less than four minutes later the plane disappeared from radar, he said. 'The only other significant thing was that at approximately 7:14 GMT a carrier came up frequently and it lasted for approximately five or six seconds. A carrier wave. In other words it is a [radio] wave which was not modulated – a carrier with indecipherable modulation – it was like a shout or a noise of some description.

'One would think the microphone was turned on,' he continued. 'At the end of that wave a noise – there was something there. Other people say it was a screech or a voice.'

Eglington said any number of scenarios could be put together on how that last transmission occurred from the aircraft. It could mean that the pilot was desperately trying to convey a message or that the microphone button could have been pressed inadvertently.

Thomas O'Connor, a plastic surgeon and deputy medical director of Cork Regional Hospital related the grim events that transpired subsequently as a major search effort was launched and bodies began arriving in Cork. It had become a city of death.

'It was something we had not experienced in the past in any of the previous major disasters,' he reported grimly. 'Out mortuary facilities were insufficient and the gymnasium had to be used,' he said. Army trucks had begun ferrying the bodies of the victims to hospital. The bodies had been labelled with numbers. The passengers had already become grim statistics.

'They were labelled and each was given a number. We in fact did use the same numbers as the numbers given by police to prevent confusion. We formed teams of a doctor, a nurse and a clerical officer to make sure we had a record in relation to each body. Each victim was medically examined and relevant information was entered on charts.'

O'Connor said arrangements had to be made for shelving in refrigeration units to accommodate the 131 bodies which had been recovered. Arrangements were then made for post-mortems to be performed by seven teams of doctors. All the bodies were subjected to X-ray. The scope of the disaster was massive, and yet the normal functions of Cork hospital had to be carried on too. The bishop had already offered the church for religious ceremonies of all kinds for relatives who were arriving. In fact several inter-denominational services were arranged. Sikhs and Hindus, Moslems and Christians, were praying together in one church. Sadly, though, it was religious strife, apparently, that had caused the deaths of their loved ones.

Professor Cuimin Doyle, head of the Histopathology Department at Cork Hospital, was the first witness to testify about the post-mortems which began on the afternoon of Monday 24 June and continued till late Thursday evening. He performed examinations on 23 bodies in that period, while fellow pathologists carried out examinations on the rest of the dead.

'My information is, and of course I myself performed post-mortems on 23 bodies, but my information is, no evidence of an explosion was found,' Doyle stated. 'There was no evidence of burning on any of the bodies. There was no evidence of which I know of fire and no obnoxious sign in so far as explosive substances were found.'

'What you are really saying is that if there was an explosive substance activated on the plane it did not touch the areas where these people were sitting?' asked the coroner.

'Yes, that's correct.'

Doyle agreed with Coroner Riordan's observation that if a bomb had exploded in the baggage compartment in the front of the aircraft, underneath the area where passengers were sitting, no evidence of the explosion would necessarily be found on the bodies of the victims – bearing in mind that a majority of those recovered had been sitting in the back of the aircraft.

Doyle was excused for the moment. Next called to the witness stand was policeman Con McGrath, who was in charge of the laborious process of identification of the bodies.

'Each body was fingerprinted by members of the Garda and each body was photographed,' McGrath told the assembly. 'Physical features such as age, sex, clothing, jewellery or any other matter for record was put on forms. The items of clothing on each body were removed and placed in a bag. That bag was numbered with the number of the body from which the clothing was removed and there was one bag per body. Likewise, items of jewellery were removed from each body and placed in a plastic bag and numbered. Each body was again photographed, this time in the nude. There was a dental examination of each body.'

When the relatives arrived, they were shown pictures, items of familiar clothing, jewellery, familiar scars. The grim process enabled police to identify all 131 bodies that had been found.

Doyle came back to the stand and described the standard medical procedure employed in post-mortems: examination of body tissue, internal organs and brain tissue, and microscopic examination of small samples of flesh taken from bodies.

One of the bodies he examined was the one labelled body number four. About this victim Doyle said, 'This body was that of a young girl, an Indian girl of about ten years. She was of average build and partly clothed. There was no evidence of burning. There was some evidence of the body having been in the water. The external examination showed there was laceration of the scalp and there were fractures of the facial bones and of the left leg, and there were other bruises and abrasions on the surface of the skin. Internally, there were multiple fragmented fractures of the nose and the skull with tearing of the membrane and of the brain itself. The lungs were damaged and there was haemorrhage.'

If that sounded like a cruel death, the court was to hear of worse cases. This child's body was one which was in relatively good condition. The terrorists should have attended the inquest to see what their handiwork had done. Another body, that of a girl of only nine years who had been sitting in the tail end of the aircraft, showed extensive dislocation of bones. For example, in this child's case, both shoulders were broken, the right leg was broken above the ankle, the skull was fractured and the spinal column destroyed. She had suffered what is described as flail injury, caused by wild rotation of the limbs after being thrown clear from the aircraft at high altitude. This meant that the tail section of the aircraft had opened up in the air.

'My understanding is that flail injury occurs when the body is thrown out of the plane. My understanding is that as the body descends it flutters like a leaf and the limbs are thrown about and clothing torn by the air,' said Doyle.

Doyle also shed light on one of the more interesting findings of the inquest. Lower portions of the bodies of many of the victims showed penetration by tiny fragments of plastic and in some cases metal. Many of these fragments had struck the bodies as though the fragments had shot upward from somewhere below the bodies. There were also flakes of paint which had penetrated the bodies of some of the victims, indicating to Coroner Riordan that the particles had been propelled by a high velocity stream of air. That too would indicate breakup of the plane or opening of its pressurized cabin while in flight.

Dr John Hogan reported a nearly incredible finding he made while examining a woman who had been five months pregnant. Unbelievably, she was alive at the time of hitting the water. She had fallen from the sky and survived – only to die of drowning.

'The other significant findings were large amounts of frothy fluid in her mouth and nostrils, and all of the air passages and the lungs were water-logged and extremely heavy,' said the doctor about this victim. 'There was water in the stomach and the uterine. The uterus contained a normal male foetus of approximately five months. The foetus was not traumatized and in my opinion death was due to drowning.'

The woman had died without even knowing that she was going to have a baby boy. The doctor said similar injuries received in a car accident could have been survivable. But the woman and her unborn child did not stand a chance in this case. Not after she hit the water, probably already unconscious.

Yet again, the topic of fragments of metal embedded in bodies came up during the course of the doctor's testimony. But again the coroner was advised that Canadian authorities had asked for forensic evidence to be kept a secret.

Barry Galvin, State Solicitor for Cork, stated that 'there is a murder investigation going on arising out of this incident – certainly not [originating] on our shore – [to determine if] there was an explosion and as a result of the explosion ... these people were murdered. So far as forensic reports are very relevant to the murder investigation it is privileged. At the request of Canadian authorities the Attorney General has been asked to keep the documents in this regard privileged.'

The key witness at the inquest was Dr Ian Hill of the UK Accident Investigation Branch. Since 1975 he has probed numerous air crashes of all kinds and is a specialist in aviation pathology.

Doctor Hill had analysed the severity of injuries suffered by the victims whose bodies had been recovered, a majority of them women and children – there were some 80 children aboard Air India 182 – and come to the conclusion that the most severe injuries had been suffered by people in the rear of the aircraft.

'It suggests to me,' the doctor said, 'that something went very wrong in the tail section because a lot of the force was transmitted to those individuals sitting there. If we look at pelvic injuries, there appears to be a concentration of them in the rear portion of the aircraft, particularly on the right-hand side. These people sustained a severe vertical loading passing up through their seats, not necessarily aircraft seats, but through their seats.'

Air India maintenance engineer Parasnuram Kaule was then called in to enlighten the coroner about the structure of a 747, including its passenger cabins, luggage holds and electronic equipment and their location.

Kaule said that the forward luggage hold, which begins

right behind the vital electronics bay of the aircraft, is about 42 feet long, 10 feet high and 12 feet wide. The aft hold is 62 feet long and shares the same dimensions of width and height, he said.

'The Main Equipment Centre of the aircraft is situated forward of the cargo/baggage hold and aft of the nose undercarriage bay,' Kaule continued. 'The main electronic centre contains much of the aircraft's electronic equipment, including two transponder sets, three VHF radio sets and the Flight Data Acquisition Unit. The two transponder antennae are located immediately below the main equipment centre. AC power for the aircraft's electronic equipment is provided by four engine-driven generators.

'There are also batteries adjacent to the flight engineer's section in the cockpit, providing emergency DC power to the captain's essential flight instruments and one VHF radio set. Emergency battery power is not available for the two transponders nor for the Flight Data Acquisition Unit. The Digital Flight Data Recorder is located adjacent to the Cockpit Voice Recorder and it is fed from the Flight Data Acquisition Unit in the main equipment centre. It monitors some 70 parameters of air frame, engine and systems behaviour and performance.'

Further, Kaule pointed out that from the electronics bay, data obtained by the Flight Data Acquisition Centre is fed through a cable below the floor level to the black boxes in the rear of the aircraft. He said that if the cable were severed recording would stop.

'Everything is gone if the Main Equipment Centre is gone?' asked Riordan.

'Yes,' replied Kaule.

'If an event occurred which would have physically knocked out all the electric power, including the communications system in the pilot's cockpit, such an event would have to occur in an area towards the front of the plane? There is nowhere else from where it could have produced such a catastrophic effect?' Riordan asked.

'No,' replied the London-based maintenance engineer.

At the end of five days of testimony, the Coroner of Cork

would receive a lecture from the Canadian government lawyer about what he could and could not do at the inquest. Mr Whitehall presented a long argument to Riordan about the restrictions he was under in terms of his jurisdiction.

The coroner shot back the protest that he was being advised to play a merely administrative role of being a registrar of the deaths. He was being told just to 'fill in a form'. Whitehall then said he didn't even see how the jury was in any position to make any recommendations in this case.

Riordan knew he was being squeezed so that he wouldn't say what was on his mind. He replied that he could see nothing wrong with narrowing down the possible causes of the accident to either structural failure or an explosive device. 'There are others but they are a bit far fetched and remote possibilities,' the coroner pointed out.

'Only when you examine the facts in detail,' argued Whitehall. 'With respect, you had no facts to examine.'

'I had the fact that it is almost certain, I think there is sufficient evidence there to be very suggestive that there was a major violent issue up there, near the nerve centre of the plane, there was a violent incident up there,' Riordan retorted.

'With respect, sir, you don't have that evidence,' said the Canadian lawyer.

'Well, it is very suggestive, I have evidence that both transponders failed, I have very strong evidence that the pilots were rendered incapable at the same time and I have very strong evidence that the navigational systems failed all at once,' Riordan shot back.

Whitehall then said that the evidence of system failures had come from a mechanic, not an expert. The coroner reminded him that the Air India man was a mechanical engineer.

Despite all the objections, the coroner was determined to speak his mind.

'I am satisfied that Section 30 [in connection with the legal powers of a coroner] is not as restrictive as Mr Whitehall has submitted. I feel the section provides authority to go behind the immediate cause of death. If it didn't I feel that the role of the coroner would merely be that of completing an adminitrative document for the purpose of registering the deaths.'

However, the coroner agreed that it would be premature for him to make a ruling over security at Canadian airports, or to conclusively point his finger at an explosive device. He had pulled his punch, despite the fact that he knew he was on the right track.

'There in the loneliness of the Atlantic Ocean lay lifeless wreckage and human bodies floating on the water,' he said. 'This was 51.03 degrees north and 12.44 west. That was about five miles away from where the airliner was when it went off the screen. One of the wonders of modern technology had been reduced to a pitiful sight.'

Regarding the technical information he had heard, the coroner continued, it appeared that all the electrical systems had failed simultaneously on this modern marvel of an aircraft. Without a moment's warning. He came to this conclusion:

'In short, if the nerve centre should fail the plane has had a serious stroke, call it what you like. The plane was almost certainly in autopilot and it was travelling at about 600 miles per hour at the time of the incident and it was in the sea five miles ahead so it must have lost altitude very, very quickly. It would do five miles in half a minute. This suggests that the first violent incident put the nerve centre completely out of order and the pilot and co-pilot were rendered incapable.'

Riordan nevertheless asked his jury to make no positive finding, and not to make a recommendation. But outside the legal confines of the coroner's court, months later, he would confide:

'You know I didn't get much cooperation in terms of calling the witnesses. Only a bomb explains everything.'

He was right, lawyer Whitehall's strenuous objections notwithstanding. The marvel of modern technology had not been reduced to pieces of flotsam by anything less than a bomb.

13

The Khola Report

While it had become quite clear from circumstantial and criminal evidence that there was a bag bomb aboard Air India 182, accident investigators still had a major task ahead of them. First, they had to map the wreckage lying on the ocean floor, at depths of 7,000 feet in some locations. Second, they had to scientifically analyse the floating wreckage recovered by various ships which had plucked bodies and debris from the surface of the Atlantic.

The floating debris recovered in the days following the Kanishka crash constituted only about five per cent of the aircraft's structure. The scientists and technical experts probing the crash had to arrive at their own conclusion, separate from the criminal evidence. The job of correlating the criminal evidence and the findings of the crash investigators was to be done by Justice Bhupinder Kirpal, a New Delhi judge appointed by the government of India to head an inquiry into the tragedy.

The task of examining the floating wreckage was left to the structure group formed by chief investigator H. S. Khola. Much of the floating wreckage was in the form of lightweight parts of the aircraft. The major items recovered included panels from the wings, spoilers which are used to slow an aircraft while in flight, engine cowlings, toilet doors, cabin floor panels, passenger seats and life vests as well as hand baggage and suitcases. Also found were landing gear, wheel-well doors and pieces of the tail section of the aircraft, such as the elevator and vital parts of the wings such as the aileron.

Initially, on 25 June, the floating wreckage was examined by explosives sabotage expert Eric Newton, who had been called out of retirement to go to Cork, where the wreckage lay prior to being flown to Bombay for further analysis. Newton was asked to submit a preliminary report with particular reference to explosive sabotage. Along with the floating debris, Newton examined the clothing of victims and the states of their bodies. Based on the study he carried out on the limited items of wreckage that were available, Newton made several preliminary observations.

Taking the scatter of the wreckage and the bodies into consideration, the condition of the limited amount of wreckage recovered indicated to him that the aircraft had broken up in flight before impact with the sea.

Detailed examination of the structural wreckage recovered did not reveal any evidence of collision with another aircraft and nothing was found suggestive of an external missile attack, the British expert concluded. He found no evidence of fire, internal or external, and no evidence of lightning striking the plane. After looking at all the available structural parts, Newton also said he could find no evidence of any significant corrosion, metal fatigue or other material defects that could have caused so sudden a break-up of the 747.

The expert said that all the fractures he had studied appeared to be consistent with overstressing and crash impact forces. Furthermore, examination of the clothing of the victims, most of them from the rear of the aircraft, showed no evidence of burning and no evidence of an explosion within the cabin of the aircraft.

Newton also examined 14 large and 29 small suitcases which were found floating after the crash. The damage sustained by the suitcases was due to impact forces rather than explosion. But Newton said that the fact that 14 large suitcases had been found meant that a luggage container had burst open, allowing the bags to escape. Examination of the lavatory doors which were found showed no evidence of explosion. Neither did the flight deck door. Newton ruled that he could find no significant evidence of explosion on the flight deck itself or in the first-class and tourist passenger cabins.

'The circumstantial evidence strongly suggests that a sudden and unexpected disaster occurred in flight,' concluded Newton.

Later, a more detailed examination of the floating debris was made by Khola, representatives of the Boeing Company and United States and Canadian experts. They examined, among other things, fan cowls of the working engines as well as the fifth, non-functional engine that had been recovered. They found that the cowling of the number 3 engine, mounted on the right wing, showed severe impact damage and found that it had small puncture marks from inside to outside. Punctures and severe damage were also found in cowling stored in the aft luggage compartment, and impact damage was found on the spoilers mounted on the right wing. Impact damage of a severe nature was noticed to other structures of the right wing, meaning that something had come out of the aircraft and hit the right wing while it was still in the air. The structure group formed by Khola found that all the cabin floor-boards that had been recovered had been detached from their fixings in an upward direction, meaning they had been blown upwards from the cargo compartment which is beneath the passenger cabin.

The major part of the wreckage, and most of the bodies of the victims, had sunk to the bottom of the Atlantic. While the French ship *Leon Thevenin* was searching for the flight data recorders, it had also made some seven video tapes of the wreckage on the ocean floor with the help of its underwater mini-submarine called the *Scarab*. On 17 July the Canadian Coast Guard vessel *John Cabot* began making a detailed map of the wreckage and prepared about 42 video tapes and 3,000 still photographs.

The tapes and photographs were studied at the Boeing plant near Seattle, Washington. The report of the structural group showed that the wreckage had been scattered over several miles on the ocean floor in an east-west direction, all of it within a radius of about 5.5 miles save for one torn suitcase which was found lying about 2 miles before the main wreckage scatter began.

While much of the forward part of the aircraft was lying within a short distance from the beginning of the wreckage trail, the portion of the aircraft extending from the forward end of the aft luggage compartment to the rear pressure bulkhead was lying scattered over a five-mile east-west swath without any concentration. The video tapes made by the *John Cabot* also showed that the forward fuselage section of the plane was lying inverted on the bottom of the sea and was broken into many pieces. The investigators found that the forward section of the aircraft, including the cockpit and the portion below the first class deck, had been totally destroyed.

The lower fuselage of the front cargo compartment was severely damaged. In certain areas, said the structure group, the main deck floor beams were also found to be severely damaged. The most significant find was that about two-thirds of the upper portion of the front cargo door, which is about nine feet by six feet, had been broken off and blown away. It was not located. The floor of the forward cargo compartment was badly mangled and smashed.

In sharp contrast, although the rear cargo compartment bottom showed an opening below its surface and the metal had curled back, the wreckage was in relatively good condition.

All four of the aircraft's engines were found detached from the wings. In two instances, the engine cases had split open.

In a report Khola prepared for Judge Kirpal, he noted: 'It appears that some opening up of the structure had occurred in the aft cargo compartment in the air. The tail portion perhaps separated before the aircraft impacted with water. This is also supported by the evidence that some passengers who were occupying seats in the passenger cabin above the aft cargo compartment had sustained flail injuries, indicating that they had been thrown clear of the aircraft in the air.

'The wreckage of the aft cargo compartment was found relatively in good condition,' Khola said. 'The bottom skin panel was curled back slightly ... the left hand side skin [of the rear fuselage] was found folded back along its length.

'The above nature of damage is not consistent with impact damage and corroborates break-up in the air of the aircraft in the aft cargo compartment area,' he said, adding, in the cautious jargon of the scientist: 'This is suggestive of some internal over-pressure in the area.'

But if Khola was saying in carefully guarded language that the aft cargo compartment had opened up in the air due to an explosion, that still didn't explain why the right wing suffered such massive damage, as though items had been flung out of the aircraft forward of the wing.

'It appears some objects had impacted the right wing and the right stabilizer in the air,' Khola said. 'This damage could have been caused by objects coming out of the aircraft from the portion forward of the right wing.'

That statement raised the intriguing question of why the forward luggage compartment door had been blown out. Khola believes contents of the front cargo hold could have been liberated into the air, hitting the right wing as the plane went into gyrations.

Another significant observation was that none of the fan blades of the aircraft showed any rotational damage, meaning that they were not operating when the plane hit the water.

Based on the evidence he had from the wreckage, video tapes, the Cockpit Voice Recorder and the Data Recorder, Khola wrote up a report indicating 20 specific findings:

* The aircraft had a current certificate of air-worthiness and was maintained according to approved schedules.
* The loading and centre of gravity of the plane were within specified limits.
* The fifth pod engine did not cause any problems in flight which could have contributed to the disaster.
* Procedure laid down by Boeing for loading the engine cowlings in the aft cargo compartment was not followed (this refers to the removal of door fittings in Toronto). But this did not contribute to the accident.
* The flight crew was appropriately licensed and experienced to operate the flight.
* Weather at the time of the accident was fair and not a factor in the crash.

* There was no lightning strike to the aircraft.
* There was no evidence of fire.
* The flight was uneventful until 7:14:01 GMT, when the aircraft disappeared from the Shannon radar screen and the flight recorders stopped recording.
* The aircraft was flying at its assigned flight level of 31,000 feet and was on its assigned track until the moment it disappeared from the radar screen.
* All the four engines were operating normally until 7:14:01 GMT.
* The aircraft's speed was sometimes six knots over the specified limit of 290 knots while carrying a fifth engine. But Boeing said this would cause no problem and did not contribute to the crash.
* There was no problem regarding the controllability of the aircraft until 7:14:01 GMT.
* No emergency was declared by the crew. Life rafts were uninflated and so were the life jackets. The seats were not in an upright position as they would be if the crew had had a chance to caution passengers about a problem and institute ditching procedures.
* The wreckage of the aircraft was scattered over a distance of five and a half miles in the direction of the flight, meaning that it broke up long before it hit the water.
* The right-hand wing inboard of the No. 3 engine and the right-hand stabilizer showed impact damage sustained in the air.
* Fan blades of the engines showed no rotational damage, indicating that the engines were not operating under power when the plane plunged into the sea.
* The wreckage found in the beginning of the wreckage trail consisted mainly of suitcases and the aft cargo compartment lower skin panels indicating that some rupture had occurred in the aft cargo compartment in the air.
* The aft portion of the aircraft had separated from the aircraft, perhaps before impacting the water.
* From the sounds recorded on the Cockpit Voice Recorder and Shannon ATC tape, it appeared that an explosion had occurred on board the aircraft at 7:14:01 GMT.

Khola had said about everything there was to say, except to point a finger at M. Singh's bag. But one aspect of the crash was still something of a mystery. Was the explosion in the rear cargo compartment or in the front one? Khola seemed to believe that it was in the aft compartment, but that didn't explain why the front cargo compartment door had been blasted out.

United States expert Paul Turner, from his analysis of the Cockpit Voice Recorder and the ATC tape, had ruled that the explosion had occurred in the front of the aircraft. Also, Eric Newton was still wondering why the transponder had died so suddenly if the explosion had occurred in the back of the aircraft. The transponder sends the signal – called the 'squawk' – that enables radar to pinpoint the position of the plane. It had blanked out at the same moment the Cockpit Voice Recorder died. That could only be explained if the explosion had been close to the Electronics Bay of the aircraft in the forward belly hold.

The debate could be settled once and for all if samples of debris could be brought to the surface. But such an operation to salvage wreckage had never been carried out at depths of 6,700 feet. Could it be done?

The Challenge of the Deep

First the sceptics said a plane lying on the bottom of the murky Atlantic would never be found. Then they said the black boxes would never be recovered. And when those things had been done, they said there was no technology available to salvage the wreckage.

They hadn't reckoned on the champion of the deep, a robot submarine called the *Scarab*. The 6,300 pound marvel of modern technology with its video eyes constantly scanning, its jets pushing it forward, and its sonar guiding it in the right direction, would prove them wrong by crossing a frontier. The frontier of a 6,700 foot depth, one from which man had never before salvaged the wreckage of an aircraft.

What is more, the *Scarab* (there are actually two in the

world today) wasn't even designed for the task it was now being assigned. The submarine, a little larger than a small truck, got its name from the original job it was supposed to do. Maintain and repair undersea telecommunication cables. Hence the name Submersible Craft for Assisting Repair and Burial.

Burial indeed. Now the little craft was being asked to open a watery grave which held the 300 or so scattered parts of the Kanishka. The computerized Scarab was first developed by Bell Telephone Laboratories in New Jersey. The designers had in mind operations at a limit of some 6,000 feet. But the job called for diving to depths of more than 6,700 feet. The Scarab would go down to that depth again, as it had done while retrieving the black boxes that provided such vital clues about the flight conditions before the crash. This time, though, the logistics were slightly different. This job involved plucking much heavier objects from the sea bed and called for a tricky operation that would require two surface ships.

The Canadian government was already paying through the nose for sending the Coast Guard ice-cutter *John Cabot*, recently converted to a cable-laying ship, to the scene of the disaster shortly after the crash. The 303 foot long and 60 foot wide vessel had a complement of some 90 crew members. The ship would have to house additional technicians for the *Scarab*, which requires at least three people to control its propulsion unit, manoeuvre its video eyes and its detachable arms and operate the onboard computer. The *Scarab* is an immensely expensive machine to charter. Another ship would be needed as well because both the deck space on the *John Cabot* and its capabilities for lifting heavy aircraft parts were inadequate for the job. For that reason, the Canadian offshore supply vessel *Kreuztrum* would also be used.

The cost, according to the lowest estimates, would range from $10 to $15 million. But the Canadian government was willing to take its share of responsibility. The order came from the Prime Minister's office in Ottawa to proceed with the search, and the *Kreuztrum* was initially hired for a 30-day period, which would later be extended for another 10 days with funds coughed up by the Indian government.

It was already autumn, and winter storms in the Atlantic are not to be sneezed at. This imposed a strict limitation on the scope of the search. The Canadian Aviation Safety Board initially dispatched investigator Art La Flamme, accredited to the Kirpal Commission inquiry to oversee the salvage job from shore in Cork. After a short stint, he would then be replaced by investigators Wally Peters, Harry Boyko and Brian Mask.

The structure group set up by Indian investigator Khola had already examined videotapes and still pictures of the wreckage. They made a decision to lift out only pieces that could shed some light on what had happened – 'a sort of a cross-section of the aircraft,' as Brian Mask explained. That was what the structure group wanted. Pieces of the forward luggage hold bottom, parts of the aft luggage hold, parts of the rear pressure bulkhead and so on.

But what do you see at depths of 6,700 feet under the ocean? What is the view through *Scarab*'s eyes?

'The bottom of the ocean is just sand, flat and clear,' said Mask. 'There is nothing down there, no sign of vegetation there at all ... pieces of metal just lying there as though wreckage of an aircraft had been laid out on a beach.

'The resolution of the photography is such that you can actually read part numbers off some of the wreckage. Sometimes the moving *Scarab* raised a dust storm on the bottom. Then it would be allowed to stand still till it settled,' Mask said, describing the burial ground of Kanishka and many of its passengers. 'There is a bunch of metal pieces lying all over the place ... some are just a tangled mess and some are sitting there as though they'd been sliced neatly.

'You see a suitcase sitting there and it looks like it's in perfect condition. Then you see metal that's all pieces which range from parts the size of small suitcases to the cockpit which is 25,000 pounds. The biggest piece was the nose section – it was just a tangled mess.'

The salvage operation would begin with controllers lowering the *Scarab* from the mother ship *John Cabot*, to which it was attached by an umbilical cord of about 10,000 feet. The robot sub would clutch the smaller objects and

152

bring them to the surface, and then a basket lowered from the *Cabot* would pluck it up. The tricky part was lifting the bigger pieces, such as a big chunk of the lower skin of the forward luggage hold. First the *Scarab* would be lowered to attach a bridle to the metal chunk and bring a line back to the surface. Then crew from the *Cabot* would attach a buoy to the line. The *Cabot* would move out of the location to make way for the *Kreuztrum*, which would move into place and use its massive crane to lift the object to the surface. A couple of intriguing parts of the aircraft, such as a forward door, were located but unfortunately dropped and never retrieved. But in all more than 23 different articles were brought to the surface.

The find included a couple of bags that really interested the structure group, especially a bag that had a tear in it. Also salvaged from the bottom was a video recorder which was broken in several places. A huge chunk of carpet from the forward cabin was brought to the surface, and then an eight-foot section of the bottom skin of the forward luggage hold.

Sitting with microscopes in the *Kreuztrum* were two metallurgists, including RCMP explosives expert Ron Madore, who would provide initial analysis and advise on the need for further salvage after viewing parts that showed damage consistent with an explosion on board. Madore had already seen the forensic evidence in Narita. On shore was Vancouver RCMP crime lab expert Sandy Beveridge.

One of the most interesting finds was a small circuit board, embedded in a bag pulled from the bottom. On first examination, it looked very similar to the timer circuit board found after the Narita blast.

The forward section of the fuselage skin, which lay underneath the forward luggage hold, had 20 holes in it, punctured from inside out, as though particles had been propelled from the inside of the luggage hold. The carpet from the cabin showed holes blasted in an upward direction from the luggage holds. Furthermore, some parts of the passenger cabin showed particles of glass embedded in panelling. Further analysis would produce still more evidence that a bomb had knocked down the Jumbo. The *Scarab*, the men from the Canadian Coast Guard and the

Canadian Aviation Safety Board had beaten the challenge of the deep.

14

The Kirpal Inquiry

No one wanted to take blame for what had happened to Air India Flight 182. That became evident as Indian Judge Bhupinder Kirpal began his official inquiry in mid-November in New Delhi to find the cause of the crash of Kanishka. The lawyers representing various parties, such as the Canadian government, Air India, Boeing and Air Canada, wanted to make sure their clients weren't saddled with the burden of what had happened.

The reasons were obvious. Multi-million dollar civil suits by the families of the victims were pending. Boeing wanted to make sure that there was no fault attributed to the pride of their aircraft manufacturing plant, the Boeing 747. Air Canada wanted to establish that the extra engine mounted by its technicians had played no role in the crash, nor had the door that had been removed and refitted at Toronto. The Canadian government lawyer, as he had done in Cork, tried to come up with all kinds of other possible causes but a bomb. Air India's lawyer, Lalit Bhasin, badgered Mountie Mike Atkinson, of the national crime intelligence unit, as though the RCMP were in charge of baggage security at Canadian airports. Atkinson was caught in the middle. The questions regarding airport security should properly have been addressed by Transport Canada.

It wasn't until Judge Kirpal put his foot down that the fault-finding stopped and the fact-finding began. It was the

RCMP which would hand Kirpal the answer to the tragedy on a platter. Sgt. Atkinson had been elected to go to New Delhi on behalf of the task force. He walked in with Canadian government lawyer Ivan Whitehall and more than 1,000 pages of affidavits the RCMP had gathered from every witness who knew anything at all about Air India Flight 182.

The Mounties had been astonishingly thorough. They presented the court with evidence covering everything from the booking of the flight reservations for M. and L. Singh to the time the flight left Toronto and Montreal. The RCMP had covered every possible angle that could be explored, even going to the extent of visiting London to interview Captain Narendra's girlfriend Valerie Evans at Heathrow Airport's Terminal 3. The RCMP also presented Kirpal with confidential criminal information which had already convinced the police in Canada that Air India had been downed by a bomb and that the tragedy was related to the Narita bombing.

The comprehensive Khola Report on the disaster was presented to Kirpal as well, with a pile of additional reports by the various investigative groups that Khola had formed. Included in the report were transcripts of Air Traffic Control conversations with the doomed plane all the way from Montreal and Gander in Canada to the last-minute conversation of the co-pilot with Shannon. The transcripts left no doubt that everything was normal with the flight until the moment of the explosion.

Other significant aspects of Khola's report were the maintenance and rust prevention programmes undertaken on the aircraft. (The report showed that minor superficial corrosion was present in the toilet area, which is common. Khola concluded that rust had played no role in the disaster.)

In the first session of the Kirpal Commission Inquiry only three witnesses were called: Mike Atkinson of the RCMP, British pathologist Ian Hill, who had already testified in Cork, and Khola, who was cross-examined at length.

Khola testified that he was given to understand that security at Canadian Airports was the responsibility of the government agency called Transport Canada, but that Air

India had a contractual agreement with Burns Security to provide secondary security checks of baggage and passengers. Whitehall suggested to Khola that the RCMP, although present at Canadian Airports, play the role of peacekeepers rather than maintainers of baggage security. Whitehall also questioned Khola about the non-functional auxiliary power unit on the aircraft. The APU provides electricity to the aircraft when on the ground, but the unit on Kanishka had not worked since the aircraft had left Bombay, and in Toronto and Montreal the plane was connected to ground power sources.

'Is it correct that the primary source of electricity of the Boeing 747 is from the four engines while the aircraft is flying?' asked Whitehall. 'If the electricity is interrupted for some reason then it is the APU which will provide the electricity while the aircraft is flying?'

'What is suggested is correct but that is not the only source of power. The alternative source is also from the batteries,' replied Khola. But he admitted that should the engine power supply be cut off, then the APU would supply power to the Cockpit Voice Recorder and the Flight Data Recorder, which do not have battery backup. However, experts agree that if the cables supplying electricity to the black boxes are severed, both sources of power are cut off simultaneously.

Whitehall also wanted to know why Captain Narendra was declared unfit to fly in May 1975 for a period of three months. Khola said that Narendra had suffered from an uncomplicated case of diabetes mellitus, but the condition had not persisted and he was allowed to return to his job. In reply to a question from Air Canada lawyer J. P. Moore, Khola said that the fifth engine mounted on the left wing by Air Canada technicians had caused no problem that had contributed to the downing of the aircraft. He also testified that the door that had been removed and refitted in Toronto to allow the loading of the fifth engine cowling had remained secure and caused no trouble. Air Canada clearly wanted to make sure no-one would blame it for the disaster.

Dr Hill, an aviation pathologist from Farnborough, England, was part of the medical group set up by Khola to study and prepare a report on the state of the bodies and any evidence that could be obtained from that aspect of the investigation. Hill testified that of the 39 per cent of the passengers of AI-182 whose bodies had been recovered, most had been seated in the aft area of the aircraft, namely zones C, D, and E – the last zone being right above the aft cargo compartment. He found that bodies of children recovered from the crash site mercifully showed much less severity of injury than those of adults.

Hill, as he had done in Cork, testified again that passengers in the rearmost section of the aircraft had suffered most of the flail injuries, caused by wild flinging of the limbs which results in dislocation of bone joints. He said the significance of the flail injuries was that the passengers were thrown out of the aircraft at altitude, again supporting the evidence of Khola's structure group that the plane opened up suddenly while in flight. He agreed that the injuries the victims had received were not indicative of an explosion in the immediate vicinity of the passengers, but added, 'By immediate vicinity I was thinking of an explosive device being in the passenger compartment itself. An explosive device in the front or the rear cargo hold would not be regarded by me to be in the immediate vicinity.'

Hill told Judge Kirpal that 25 of the bodies he had examined showed signs of decompression. Seven of the victims who showed these signs were children. He said passengers seated on the right side of the aircraft showed more evidence of decompression, and that it would take less than a second for a person to exhibit such signs of decompression. Hill said the evidence of decompression injury is shown by extensive damage within the lung tissue and is generally more severe if the person is sitting close to the area where the decompression of the pressurized cabin occurs. Hill said a majority of the passengers probably died prior to hitting the water but at least four passengers were apparently still alive when they hit the water. Hill also stated that from the injuries he had seen, if a bomb was aboard the plane and was carried in one of the cargo holds, it was more likely to have been in the rear hold because of the massive upward vertical forces that had hit passengers sitting at the back of the plane. The only obvious conclusion to be drawn from Hill's testimony, which had varied little from what he had said during the Cork inquest, was that an explosion definitely had not occurred within the aircraft's cabin.

The man of the moment at the inquiry was the Mountie. He went into great detail about the workings of the airports at Vancouver, Toronto and Montreal and made it categorically clear that M. Singh had never boarded his flight out of Vancouver for Toronto, although he had made a fuss to get his bag interlined with Air India. He also said that the German airline Lufthansa and Air India were the only ones who X-rayed their baggage on a continuing basis. He further noted that Air India was the only airline he knew of which used a security number system to make sure all passengers who had checked in had boarded.

Atkinson was then questioned by R.K. Anand, for the government of India, about the system employed by Canadian Pacific Airlines in Vancouver where the two killer bombs had been boarded.

'You had indicated that the baggage on CP 60 [the flight which carried M. Singh's bag to AI-182] was not X-rayed. Can you tell us as to whether the baggage on CP 003 [which carried the bag bomb to Narita] was X-rayed or not?'

'I am unaware of it,' the policeman replied.

'I suggest to you that as of June 22, 1985 there was no system of having an X-ray machine in respect of the check-in baggage at Vancouver. Is that correct?'

'There were X-ray machines at Vancouver but I don't believe that they were in the baggage area.'

'It is suggested to you that if the security number system was used on Flight 60, then the baggage of M. Singh could not have gone into the aircraft. Is that correct?'

'If the full security number system with respect to baggage and passengers was used on CP 60, then the baggage should not have gone on the aircraft,' replied Atkinson.

'I suggest to you that if the security number system had been used by CP Air for its international flights, especially CP 003 [to Narita], the baggage of L. Singh would not have gone into the aircraft,' Anand said.

'If L. Singh had travelled, then the baggage would have gone onto the aircraft,' replied Atkinson, 'but if L. Singh had not travelled then the baggage would not have gone to the aircraft.'

'Did your investigation reveal that CP Flights 60 and 003 were booked and that the tickets were purchased at the same time?'

'... The tickets for L. Singh and M. Singh were booked at the same time and picked up at the same time,' said the Mountie.

'Is it correct that the contact telephone number given in respect of L. Singh and M. Singh was the same?'

'I believe it so,' said the RCMP officer. The number, of course, was the Vancouver number 437-3216, belonging formerly to the moderate Sikh, Hardial Singh Johal.

'If there is a confirmed ticket from Vancouver to Toronto and a person is wait-listed from Toronto onward,' asked the lawyer, 'would the CP agent accept the baggage to Delhi at the time of check-in at Vancouver [As Jeannie Adams had done at Vancouver Airport with M. Singh's bag]?'

'I don't believe that would be so,' replied the officer. Atkinson also agreed that a passenger wait-listed on Air India in Toronto would have to check in at the Air India counter with his luggage.

'On June 22, 1985, was there any procedure in Vancouver Airport to check the passport at the time of checking in passengers who were to be interlined for an international flight?' pressed Anand, in apparent reference to the fact that despite regulations that require agents to check passports for passengers flying on international flights, no one remembers if anybody asked M. Singh or L. Singh to produce a passport in Vancouver. The requirement is clearly printed on all CP Air passenger brochures and time-tables.

Atkinson replied, 'I do not know.'

'At Vancouver Airport on June 22, was there any procedure for counting the number of passengers who had actually boarded the aircraft and comparing it with the number of passengers who had checked in?'

'I am not aware of any procedure,' said the officer.

Then it was Bhasin's turn to grill the Mountie. The Air India lawyer suggested to him 'that keeping in view the combination of various measures taken by Air India in consultation with the RCMP, Air India had the best security system as compared to other airlines as of June 22.'

'All I can say is that Air India had security for their flights and it was more secure than for other airlines,' was Atkinson's response.

'Is it correct that Air India was the only airline which had names and address tags on all checked baggage?'

'I am not aware of that.'

'Is it a fact that Air India was the only airline which had cabin security checks before the boarding of the cabin crew?'

'No, I am not aware of that.'

'Are you aware that Air India was the only airline at that point in time to have 24-hour cargo pooling?'

'No, I am not aware of that.'

'Are you aware that Air India is the only airline which had pantry and food uplift security checks?'

'No, I am not aware of that,' replied the Mountie, taking the heat for the Disneyland security atmosphere at Canadian airports prior to the Air India and Narita blasts.

'Are you aware that Air India made a request to the RCMP in the month of January 1985 that the services of a dog should be provided for sniffing the baggage, and this request was turned down?' Anand asked.

'I am aware that there was a meeting between Air India personnel and the RCMP,' said Atkinson. 'I don't know what was discussed.'

When the first session of the Kirpal Inquiry adjourned, the whole affair had again boiled down to M. Singh and L. Singh. And their deadly bags.

The second session of the Kirpal Commission, which opened on 22 January 1986, would finally bear fruit. A Canadian Aviation Safety Board officer, Bernard Caiger, found himself caged by Boeing lawyer Steven Bell. Caiger admitted finally that he had 'briefly' seen a report that showed that a bomb in the forward luggage hold had caused the tragedy.

'I am not sure whether it says a bomb was the cause or the probable cause,' he hedged. Pressed further, he said that he had seen no other cause listed. At last, somebody was finally talking about the foregone conclusion the Mounties had reached on Day One after the crash.

Later, Canadian officials did a considerable amount of backtracking. But Caiger had said his analysis of the Cockpit Voice Recorder had shown a 'bang' when the recorder stopped working and the jet disintegrated. That was nothing new either. Paul Turner, of the US National Transportation Safety Board, had already said a long time ago that an explosion occurred close to the cockpit.

Two experts, however, one from Boeing and one from India, made it plain in their testimony at the Kirpal hearings that they believed there had been a bomb on Air India Flight 182 – although they gave different versions of where the device was located.

Boeing investigator Harold Piper testified that he felt the explosive device must have been located in the back of the aircraft, as Khola had concluded, because of the spread of the wreckage of the tail end over a wide area.

'My opinion is that there was an explosive device in the aft end of the airplane,' Piper said. 'From the wide dispersion [of the tail end] it was clear that it had broken into many pieces.'

Piper said he had reached his conclusion from studying the wreckage distribution on the sea bed. But S. N. Seshadri, the man who analysed the tape from the Cockpit Voice Recorder, concluded that the explosion was in the front end, probably about 10 to 13 metres behind the cockpit, in the forward luggage hold. Shrapnel and holes found in the wreckage of the forward hold tend to back up that belief, another Indian scientist testified.

A British expert, however, said he had failed to find any evidence of a bomb from his analysis of the recording. He agreed, though, that his findings didn't rule out a bomb.

There were lots of reasons why so many experts had been dragging their feet on the bomb question at the Kirpal inquiry. Especially those representing Canadian government bodies. The basic reason was that admitting that a bomb had downed the jet would leave the government wide open to lawsuits.

Nevertheless, Kirpal already knew the answer to what had happened to the Air India jet. When questioned by a reporter weeks before the inquiry while looking at some of the wreckage, the judge smiled when asked if he would be able to make up his mind. He knew the answer already. He nodded his head and smiled. Everybody knows. But nobody wants to be the first to say it.

15

Now They Admit It . . . Almost

On 27 January 1986, the Canadian Aviation Safety Board finally made public a comprehensive report of its findings on the Air India disaster. The detailed report began firstly with the now all too familiar scenario of L. Singh and M. Singh. But there were other more detailed findings on various aspects of the disintegration of AI-182, the Cockpit Voice Recorder analyses and the wreckage found floating off the coast of Wales, England and Ireland as well as the salvage brought to the surface by the *Scarab*.

The report, released at the Kirpal inquiry, which would later cause heated debate between Canadian lawyer Ivan Whitehall and those representing Air Canada and Air India, stated that the National Research Council of Canada had carried out an analysis of the Cockpit Voice Recorder as well as the Shannon Air Traffic Control Tapes.

'From the CVR and the DFDR (data recorder), AI-182 was proceeding normally en route from Montreal to London at an altitude of 31,000 feet and an indicated airspeed of 296 knots when the cockpit area microphone detected a sudden loud sound. The sound continued for about 0.6 seconds and then almost immediately, the line from the cockpit area microphone to the Cockpit Voice Recorder at the rear of the pressure cabin was most probably broken,' said the report.

'The initial waveform of the cockpit area microphone signal is not consistent with the sharp pressure rise expected with detonation of an explosive device close to the flight deck,' the report said. But because of the various ways in which sound from such an explosive device can travel to the cockpit area, the presence of an explosive cannot be ruled out either.

Within a split second of the sudden sound picked up by the Cockpit Voice Recorder, the Shannon ATC tape recorded unusual sounds from the stricken airliner. The sounds lasted for nearly 5.4 seconds.

'Listening to the sounds, it also appeared that a human cry occurred near the end of the recording. Spectral analysis of these sounds and comparison with voice limitations revealed that the recorded sounds do not contain all the pitch harmonic frequencies normally associated with voice sounds,' the report continued.

The CASB report said the aircraft was restricted to altitudes below 35,200 feet because of the carriage of the extra engine and speeds of less than 290 knots. The data recorder showed that during the last 27 minutes of flight, the speed was adjusted several times by the cockpit crew. At one time the speed increased by over nine knots above the limit. At 07.1: four minutes before the disaster, the speed was again increased after being slowed nine minutes earlier. The report agreed with Boeing's findings that these surges would not have contributed in any way to the accident. Boeing had reported earlier that even if the speed limit was exceeded greatly the only adverse effect would be vibration of the aircraft. Furthermore, the crew was aware of the speed limitations and had asked for Oceanic clearance when leaving Canadian territory at a reduced speed.

The CASB document shows that the research council concluded that both the CVR and the DFDR stopped recording simultaneously. As well, when the signal from the transponder vanished at Shannon radar, it did not show any variation in altitude.

The report notes that the Accident Investigation Branch of the UK also came to the conclusion that a high explosive had not gone off near the flight deck. The AIB of UK had also failed to decipher the sounds heard on the Cockpit Voice

Recorder and ruled they were similar to sounds previously heard when explosive decompression had occurred on a DC-10 aircraft. Explosive decompression meaning rapid and sudden loss of pressure within the cabin. Eric Newton, the aviation sabotage expert, had called this sound a 'decompression roar'. The AIB said it could not determine the origin of the explosive decompression.

Indian scientists, based at the Bhabha Atomic Research Centre (BARC) came to a different conclusion. The cockpit area mike showed a rise in volume from the normal sounds in the cockpit in about 45 milli-seconds. The rise was about 18.5 decibels higher than normal sound in the cockpit. The Indian scientists compared the sound with an explosion which had caused the crash of an Indian Airlines jet, a Boeing 737, and found that the sound signal had risen from normal cockpit ambient in about eight milliseconds. The explosion in that case had occurred about eight feet from the microphone. The scientists considered that relevant, therefore they concluded that the explosion on Air India 182 had occurred about 40 to 50 feet from the cockpit microphone, which is the site of the beginning of the forward luggage hold.

The CASB report then included the findings of aviation explosives expert V. J. Clancy, of Boeing, who had closely examined floating debris recovered from the crash site.

Clancy's findings showed:

* A foam-backed floor panel had a small number of perforations.
* One of the lavatory doors had embedded in it a number of shards of mirror normally fitted in aircraft toilets. Clancy, however, said he could not put much relevance on this finding.
* Three oxygen cylinders which were stowed in the forward cargo compartment were recovered. One showed a dent made by the impact of an object measuring about one to two centimetres. The depression was tiny and could have been caused by high velocity particles hitting the bottle.
* A red suitcase with blue inner lining was recovered from the ocean. Clancy noted the lining was tattered in the same fashion as a suitcase recovered from an Angola aircraft which

had been subjected to a bomb.

* A wooden spares box, normally stored at the back but it could have been in the front on AI-182, showed burns. It had been exposed to burning for about four minutes.

* Two pieces of the cover of an over-head locker, probably from the front of the aircraft, were partially damaged and blackened by fire.

The Canadian Aviation Safety Board studied the wreckage too and found that the number 4 engine had a series of marks as though it had been hit by a turbine blade from the number 3 engine also mounted on the right wing. An upper deck cabinet showed a rounded dent which was not consistent with impact damage. The CASB said it was probably caused by an explosive shockwave generated below the cabin. There was also blackening on the bottom of some seat cushions. Investigators said the charring was similar to that which would be caused by an explosive device.

CASB said the first suitcase found at the beginning of the wreckage trail, the one which came out of the aircraft first as the plane went down, was not recovered. The suitcase was seen with clothes protruding from a tear. The CASB gave no reason for the failure to recover the suitcase.

Another note of significance was that the strut that held up the number 3 engine on the right wing was found a long distance away from the main scatter of engine wreckage. In addition, one of the working engines was lying 0.5 nautical miles away from the engine scatter.

The video tapes showed that a section consisting of the cockpit, first class and electronics bay was so badly mangled that neither the cockpit nor the electronic bay were distinguishable. Portions of the forward cargo hold, the upper crown of the aircraft and the upper deck passenger area were found close to the forward nose section of the aircraft. Scattered nearby were suitcases and badly damaged luggage containers.

All cargo doors were found intact and attached to the fuselage except for the forward cargo door measuring about nine feet by six feet. The upper two-thirds of the door had been ripped out, while the portion that remained attached to the fuselage was badly frayed. The door was located but dropped accidentally during the salvage operation, and this most crucial part of the wreckage has not been located since.

As observed earlier by Khola, the tail section of the aircraft was scattered over a wide area, indicating break-up and disintegration at altitude.

One of the most significant items recovered during the salvage operation was the right hand side forward fuselage section just behind the front cargo door. There were numerous outward holes in this section, and when fished out along with it came hundreds of tiny fragments and medium-sized pieces which had pierced the skin. Scientists at Bhabha Atomic Research Centre, the Indian National Aeronautical Laboratory and the explosives research laboratory conducted metallurgical tests on this item of the fuselage skin.

They found some of the curling of the skin was indicative of a shock-wave effect, that the large number of fragments of non-brittle aluminium was indicative of explosive forces and that the punctures and outward petalling of the metal were all also indicative of an explosion.

The second item with remarkable bomb clues was the lower skin of the forward cargo hold with about 20 holes in it. Boeing expert Clancy noted that some of the holes were made by high-velocity particles propelled from inside the aircraft such as those produced by an explosion. Part of the panel was blackened by fire.

This area, too, came with hundreds of tiny fragments, again indicative of an explosion. One hole in the skin was described as something akin to a bullet-hole. While Clancy was guarded, the Indians were more open about what these features indicated. The Indians that concluded there was explosive loading in this part of the aircraft while Clancy said it showed the 'possibility' of an explosive.

The CASB report said examination of the wreckage indicated that the right wing and the number 3 engine suffered impact damage in the air. Further, the forward cargo door was broken and blown off from the aircraft and may have struck the wing and the engine. 'The damage to the door and the fuselage skin near the door appeared to have been caused by an outward force and the fracture surfaces of the door appeared to be badly frayed.

'A failure of this door would explain the impact damage to the right wing areas. The door failing as an initial event would cause an explosive decompression leading to downward forces on the cabin floor as a result of the differences in pressure between the upper and lower portions of the aircraft. However, examination showed that the cabin floor panels separated from the support structure in an upward direction. Also, passenger seats recovered showed they had been subjected to an upward force from below. They showed that seats in the rear had their back legs buckled and the seats toward the front had both front and back legs buckled. This indicates the vertical force was greater at the front than the rear of the aircraft.'

It didn't say in so many words, but what the CASB was saying was that the door being blown away as a primary event would not have caused the upward force on the cabin floor boards. The door being blown away and the upward force could only mean one thing. That an explosion blew away the door and caused a vertical load from the bottom to the top.

Again, very cautiously, the board made this remark in its report: 'There is a considerable amount of circumstantial evidence and other evidence that an explosive device caused the occurrence. The evidence indicates that if there was an explosion, it most likely occurred in the forward cargo hold, not the passenger and flight deck areas or exterior to the fuselage. Although an explosive device could have been placed in a cargo hold in a number of ways, the available evidence points to the events involving the checked baggage of M. and L. Singh in Vancouver.'

So now they had finally blurted it out. Where do you put the blame?

'Canadian security arrangements in place prior to 23 June 1985 met or exceeded the international requirements for civil air transportation. However, before this date, the emphasis was on preventing the boarding of weapons including explosives in hand-luggage,' said the authors of the report.

Hijackings were considered the primary threat to airliners. But history has no shortage of examples of bag bombs.

'In Canada, the Department of Transport (Transport Canada) is responsible for establishing overall security standards for airports and airlines and for the provision of certain security equipment and facilities at airports. By regulation air carriers are responsible for applying security standards for passengers, for baggage and cargo, and for ensuring security within individual aircraft. The RCMP provides physical security and responds to criminal incidents.'

The report adds that airlines contract the services of private security firms to undertake security checks. But 'Transport Canada has established certain standards required for licenced security guards such as the completion of the Transport Canada passenger inspection training program and annual refresher training.

'A significant number of the security guards [who were on duty X-raying and testing for explosives when Air India 182 left Toronto] did not meet the criteria with respect to completion of the training program and refresher training. In addition, the criteria do not require training for the screening of cargo and checked baggage.'

The report did not discuss the question of whether Transport Canada did not enforce the criteria which it set.

Instead, the report states that annex 17 of the International Civil Aviation Organization, to which Canada is a signatory recommends: 'That contracting states establish the necessary procedures to prevent the unauthorized introduction of explosives or incendiary devices in baggage or cargo intended to be carried on board aircraft.'

The report said there were provisions in place to prevent unauthorized boarding of cargo or bags on board aircraft.

'However, if someone were to purchase a ticket, check in baggage and not board the aircraft, the baggage would in all likelihood have been authorized by the airline to be placed on board the aircraft. Therefore, it was possible to interline baggage unaccompanied and this explains how a suitcase was interlined to AI-182 from CP 60. It is not the normal practice of airlines to interline baggage if there is not a confirmed reservation to the destination. In this case, the ticket agent allowed the suitcase to proceed; however, if there had been a confirmed reservation, the suitcase would have been interlined unaccompanied without question.'

The report had forgotten one thing, though. How would that suitcase have gone to Toronto if a count had been taken of those who had checked in and boarded at Vancouver? Should not the suitcase of a no-show passenger have been removed? Would not X-raying at Vancouver airport have prevented L. Singh's bag from going to Narita with its killer bomb?

The CASB pulled its punch when making a final pronouncement on the death of AI-182. It said the initial event was probably an explosion in the front hold. 'This evidence is not conclusive. However, the evidence does not support any other conclusion.' It was time somebody said the obvious, without the ifs and buts.

16

And Now The Truth

The experts and the lawyers can argue until they are blue in the face about whether Air India Flight 182 was downed by a bomb. But the RCMP already know the answer. Just as they did on Day One of the probe. But now they've taken the investigation further. As this is being written, they know most of the men responsible too. They know who is responsible for the bloodiest-ever case of aviation sabotage.

The answers didn't come overnight. It has proven to be the costliest RCMP investigation in history. But the Mounties know that a bomb concealed in a bag checked in the name of M. Singh and boarded on CP Air Flight 60 in Vancouver destroyed the Boeing 747. They even know the probable explosive power of the bomb carried in that suitcase which ended the lives of 329 people. It is a case of murder. The murder of 329 people aboard Air India 182 and two more by a similar bomb at Narita. Pure and simple.

The fact of the matter is that experts cannot, as Eric Newton so validly pointed out in his article in the Journal of Aviation Safety, consider any aspect of the crash in isolation from other factors and come to a valid conclusion. That has been the whole problem with the Kirpal inquiry from the outset. One expert would just analyse the sounds recorded on the Cockpit Voice Recorder. Another would take a look at the Air Traffic Control tapes. Yet another would look at the bodies that were found. And each one would come back and say there was no evidence of a bomb.

What is needed in an investigation like this is correlation of *all* the facts relating to the case, from its beginnings in Vancouver to the sudden and tragic end of the aircraft at 7:14:01 GMT. Newton says it often happens in similar cases that actual bomb fragments are difficult clues to find. In this case those fragments are lying at the bottom of the sea, lost forever. Therefore, the only valid conclusions can come from circumstantial evidence, both the evidence of criminal activity and the evidence of the sudden and dramatic disappearance of that flight close to its destination, Heathrow Airport, when nothing had been wrong with the aircraft throughout the whole flight.

Take one example of expert testimony. Dr Hill can only say that there was no evidence of a bomb in the bodies he examined. However, he got samples from only 131 bodies and one more that was recovered during the salvage operation. That does not rule out the odd-man theory – that somewhere in the wreckage of the aircraft under the murky waters of the Atlantic is a body that does contain evidence of an explosion. Hill can only say he did not find any evidence of a bomb on the people he examined. At the same time, he can't rule it out either. There is no doubt anyway that the bomb was not located in the passenger cabin. That was ascertained for a fact through other evidence.

Cockpit Voice Recorder experts tried to analyse the sound of a bang and came to different conclusions about what that sound was. On its own, that sound would not be sufficient to prove whether or not the sound was made by a bomb. On the other hand, no one can say that the sound was *not* made by a bomb. All that can be said, and all that needs to be said, is that it is the sound of an explosion.

The salvage operation showed unique holes and blast damage that were consistent with an explosive device. Holes are one of the best indications of this. Secondly, at least one piece of the interior panel of the aircraft which washed ashore showed signs of blisters, as though it had been burned. Furthermore, a carpet from the cabin shows holes made in an upward direction from the bottom of the forward luggage hold. Particles of debris were flying inside the cabin, as evidence in Cork showed. Metal and plastic particles had become embedded in the bodies of the victims. That could only have been caused if a violent airstream entered the fuselage from some point in the aircraft, likely from the front. Other evidence shows that the forward cargo door, measuring 9 feet by 6 feet, was torn away – the upper two-thirds of it – not torn from the hinges, but ripped out. The hole this left in the aircraft would have been about 30 sq.ft. More than enough to cause massive structural failure, as Newton said.

That door didn't rip itself apart. The door gave way to a massive force pushing against it from the inside of the front luggage hold. The container loading pattern in the aircraft's forward luggage hold, which is slap against the electronics bay, shows that two containers on the port side were carrying bags bound for New Delhi from Toronto, while one container, the forward right side container, lying against the door of the cargo hold, was empty.

M. Singh's bag would not have arrived at Terminal Two of Toronto airport until late in the process of loading. Therefore it would have been among the last bags loaded onto the aircraft. It too was marked with the destination New Delhi. It could have been, and probably was, in one of the two containers lying close to the electronics bay. A massive explosion caused by more than eight sticks of dynamite in either one of those two containers would blast the empty container right through the door, breaking it two-thirds of the way down. This is confirmed by the fact that the right wing and an engine mounted on it – the wing is just behind the cargo door – showed impact damage in the air, as though items had been thrown from the front of the aircraft. The experts cannot deny the scenario that a blast in the forward luggage hold flung containers and bags out of the cargo door and against the wing.

Such an opening in the fuselage would cause massive explosive decompression, the bomb lifting the floorboards of the cabin in an upward direction, as the evidence showed, and causing a stream of air carrying debris of metal and plastic to hit passengers as the jet travelled at approximately 600 miles per hour. The opening would also result in the jet, with its right wing damaged already, suddenly pitching nose down. This would cause a violent upward jerk on passengers sitting in the back of the aircraft – vertical loading, as the experts called it.

All of this would happen rapidly, within split seconds. The aircraft would start disintegrating from all sides. Further evidence of the bomb being in the forward luggage container is the fact that all electronic equipment stopped working simultaneously. An explosion in the back would not have caused this. It would have allowed the transponder to keep going for a while, at least. That it did not is symptomatic of sudden failure of the Main Equipment Centre, the plane's vital nerve centre. However, all these arguments, as Eric Newton says, are academic and an aircraft is doomed no matter where a bomb is placed if the explosive is of sufficient power. Well, it was in this case. The Mounties know, having taken their clue from the power of the Narita bomb, which they have no doubt was similar to the Air India bomb.

For the sake of argument, even supposing that no physical evidence of a bomb was ever found, can the experts explain away the circumstantial evidence compiled by the RCMP? Can they explain the tickets of L. Singh and M. Singh? Can they explain why both tickets were bought together, reserved together? Can they explain why the telephone contact numbers for both were the same? Explain why the man whose number was given doesn't know them? Why both men checked in their bags and then didn't fly?

Can they explain why both flights on which the men did not turn up ran into trouble? Why the Narita bomb exploded *exactly* one hour before Air India 182 went down? Can they explain why L. Singh was to connect with Air India in Narita while M. Singh was to connect with Air India in Toronto, if this fact was not related to the two tragedies? Can they explain why M. Singh made a fuss to get his bag interlined onto Air India 182 yet didn't get on the flight?

Finally, can they explain why the Boeing 747 vanished so suddenly while flying routinely at 31,000 feet with absolutely nothing wrong aboard the aircraft? Can they?

Those are all the factors that have to be taken into account to come to the conclusion the RCMP have reached. They already know not only that there was a bomb, but who placed it. They have evidence that a member of a Sikh fundamentalist organization persuaded a man in British Columbia to part with more than enough sticks of dynamite to make two bombs of the size that exploded in Narita, with some sticks left over. The man had bought, in the spring of 1985, sufficient quantities of dynamite, black powder and blasting caps to sink a battleship.

Police know both the buyer and the seller. All the RCMP had to do was some simple mathematics after establishing the power of the Narita bomb, deducting the number of dynamite sticks used from the total number purchased. So where did the missing sticks go? Japanese police have told the RCMP, and the Mounties have carried out their own tests to prove it, that had the bomb sent to Narita gone off aboard CP 003 it would have proven fatal for the aircraft and its 390 passengers.

In Air India's case two unfortunate things happened. First, the aircraft was delayed at Toronto airport, and secondly, the bag bomb by a strange quirk of fate ended up in a position on the plane that made the aircraft particularly vulnerable. Had it not been delayed, the Air India jet would have made it to Heathrow and beaten the timer. But no-one can say what havoc would have been created at Heathrow if the explosion had happened during refuelling.

The RCMP know that two bombs were made and they know who made them. The man who made the bombs knows that the Mounties know he did it. They also know who purchased the container of the Japanese bomb, the Sanyo FMT 611K tuner that blew up in Narita. And they are even aware of the fact that on the morning M. Singh and L. Singh were to check in the bombs at Vancouver Airport there was a sinister meeting in a house near the city.

It was time to arm the bombs. The terrorists wouldn't have wanted to leave the timer ticking while the infernal devices sat around in their own home. But now the countdown to disaster had begun.

The police know that the man who made the bombs had travelled to Vancouver the night of 21 June, the night before the bombs were placed. And after M. Singh succeeded in checking in the bomb destined for Air India 182, the man who made the deadly devices found that one of the parts of the bomb destined for Narita (CP 003 did not leave Vancouver until 1:15 p.m., while CP 60 left at 9:00 a.m.) had suddenly packed up. The man who made the bombs then had to make a quick trip to a Vancouver area store (name withheld) to replace the defective part. Police know what part he needed.

There is one little step that, at the time this is being written, still remains to be taken. Finding two more people. The men who actually checked in the baggage. The question is only of the real identities of the two men. They were probably imported from outside Vancouver just for one day to check in the bags. Both were clean shaven, to leave the impression that hunted terrorists Lal Singh and Ammand Singh, fugitives from the FBI, had checked in the bags. But the rest of the group of the Dirty Dozen are all in the bag. The ringleaders have been identified. And the terrorists know that too. It's only a matter of time before Inspector Hoadley and the RCMP members working with him see how many fish are on the hook.

The Canadian government took drastic steps soon after the Air India tragedy to plug gaping holes in security at the airports in the country. It revoked an earlier decision to remove police patrols from the airports and ordered X-ray equipment in to check bags before they leave Canada. The barn door was shut, shortly after the horse bolted.

Interlining of bags is not permitted at all now when the bags are connecting onto international flights. Passengers are now required to reboard their baggage at the final point of departure from Canada, where proper counts can be taken to ensure that all passengers have boarded the aircraft. And when a passenger goes missing, everyone is forced out onto the tarmac to take stock of their baggage, so that the bag of a missing passenger will not fly without him. That of course will not eliminate a suicidal terrorist. But it does eliminate cowardly terrorists like M. and L. Singh, who considered their own lives so precious that they did not board the aircrafts with their bags. M. Singh was actually seen lifting his bag gingerly as he stood in the lineup at Vancouver Airport on Saturday 22 June. The man behind him said he would just push his own bag along with his foot. But M. Singh, whoever he is, would pick his up and set it down gently every time the line inched forward.

Two days after the Air India disaster X-ray machines suddenly appeared at Vancouver Airport. There were no checks made of checked-in baggage previously, but now all bags are X-rayed before being put on board. In Toronto, similarly, passengers arriving at Terminal One now must take their own luggage to Terminal Two to ensure that there will be no more M. Singhs and L. Singhs taking advantage of security loopholes.

Tragically, though, the twin disasters could have been averted if Canadian authorities had listened to the Canadian Airline Flight Attendants Association in 1983. Larry Le Blanc, national president, made a series of recommendations to ensure security of passengers. Among those recommendations was this memo to a government task force:

'In many parts of Europe, all checked baggage is screened by X-ray before it is loaded into the aircraft. We believe this practice should be mandatory at all Canadian airports. Furthermore, passengers should not be allowed to check their own baggage. Procedures should be implemented to prevent individuals other than airport employees from placing a suitcase on the baggage conveyor belt.'

The reason for not implementing those changes in 1983 was money. But the total cost of investigating the Air India and Narita bombs is likely to exceed 50 million dollars to the Canadian taxpayer. Furthermore, civil litigation and inquiries in India, Japan, England and Ireland will probably raise the total tab, including the insurance payments for the Air India jet, to the order of $400 million or more.

Money aside, those Canadians and Indians and Japanese

who were so brutally killed would probably still be alive if Canadian airports had been better secured.

And Vancouver wouldn't have earned the black mark of being the city where the world's worst terrorist attack of its kind was launched.

PART THREE

THE PUNJAB:
CAUSE AND EFFECT

1

Seeds of Trouble
in the Punjab

'The Sikhs are a brave people. They will know how to safeguard their rights by the exercise of arms if it should ever come to that . . .'

- Mahatma Gandhi, March 1931.

The great soul of India was speaking to a large Sikh congregation as his movement to spear-head India out of the British Raj gained momentum. But even as he spoke that day, Mahatma Ghandi knew of the peril India would face in the coming years as he slowly edged the jewel of the British Crown towards freedom. That peril was the conflicting interests of multitudes of different cultures and religions which formed, at that time, the united nation of India under the Raj. And the Mahatma was aware that once the British had left India, the polarization of religions and cultures would tear at the country like at no other time since the whip had been held by London.

Ghandi's vision of a free India was a country where Hindus, Moslems, Christians and Sikhs could co-exist without friction. A secular country where religion would play no part in the nationhood of the infant nation-state just finding its feet.

But the Raj had already made plans for splitting the country on the basis of religion. At the time the partition of India was being contemplated, the option had been made available to the Sikhs to go their separate way if they chose. But Sikh leaders of the time opted to stay within India. Gandhi, already appalled at the thought of his country being split along religious lines, was relieved that Sikhs, the third largest power-block, did not demand a separate state of their own. That was something Gandhi could not allow to happen. Therefore, when he spoke on that March day in 1931 to a congregation of Sikhs, Gandhi, the man who advocated peace, was assuring the Sikhs that their sword would be the great equalizer should the Congress party double-cross them following the Independence of India.

It was clearly recognized by the leader of the new state of India, Jawaharlal Nehru, that Sikhs deserved special rights within a united India. But successive Indian administrations dragged their feet when it came to keeping the promise of the father of the nation that had been made in 1931 and reiterated several times later. As the doctrine of secularism matured within the administrations in New Delhi it was becoming more and more clear that giving special status and self-governing power to one of India's numerous cultures and religions would erode the ability of the central government to keep the country together. However, the central government finally bowed to pressure and formed the state of the Punjab in 1966, although leaving out pockets of Punjabi-speaking people in a neighbouring state called Haryana which was predominantly Hindi speaking. The formation of Indian states was along linguistic lines generally.

The Sikhs, whose military prowess is admired around the world, would continue to exert profound loyalty to their country and make sacrifices for India far out of proportion to their numbers even as their leadership pressed religious and economic demands. However, the balancing effect of a large segment of Moslem population having gone with the creation of Pakistan, Sikh religious leaders began fearing that their faith would slowly be eroded in a country which has a massive Hindu majority. In those fears lie the answer to the recent turmoil in the Punjab, which has shaken India right down to its foundations. And in the vicious circle of events in the Punjab lies also the answer to why 331 people died in the bombing of Air India Flight 182 and the blast at Narita.

In the beginning, though, Sikh agitation in the Punjab centred on greater regional economic power. The basic demands in the beginning were essentially religio-economic concessions, until extreme fundamentalists exploited the situation and inflamed religious feelings to plant the seeds of violent protest.

Sikhism is one of the youngest and most noble of the religions of the world, having been given momentum by the tenth Guru Gobind Singh in 1699. The religion came into existence from both Moslem and Hindu elements embracing the new faith of the *Khalsa*, the pure ones. Gobind Singh gave the Sikhs a new identity, galvanizing them into a people enjoined with the task of sharing with one's fellow-man, exerting manliness and achieving spiritual and temporal supremacy. He ordered his male followers to bear the middle name Singh, meaning the Lion, and the womenfolk to bear the middle name of Kaur, meaning princess.

There were also other tenets, such as the five basic principles of wearing unshorn hair and beard, a metal bracelet on the hand, special shorts, a *kirpan* (small ceremonial knife) and carrying a comb. The baptized Sikh is forbidden to touch alcohol and tobacco and by the nature of his religion is enjoined to be kind to those who seek refuge with him. That is why, on any given day, a stranger can go to a Sikh temple in any country and find refuge and food.

The newly-formed religion also taught valour, which has been amply demonstrated by the martial prowess of Sikhs both while fighting against the British Empire as it set foot in New Delhi and while fighting on the side of the colonial master in the World Wars. They proved their battle-hardiness in India's three wars against Pakistan too. Less than 100 years after Gobind Singh galvanized the Sikhs, they conquered Lahore and threw out the Afghan ruler, proclaiming their own king, Ranjit Singh. The Kingdom ruled for over 50 years across territory that included Lahore, which is now in Pakistan, and stretched north to the Khyber Pass, including the areas of Jammu and Kashmir.

However, the Kingdom of Punjab was overtaken by the superior armour of General Dalhousie after the second Anglo-Sikh war in 1849. The Khalsa Raj had ended and the British Raj had begun. The Sikh royal heir, Duleep Singh, was exiled to England at the age of 15, lest this boy grew to become another Lion of the Punjab. Duleep Singh died in England at the age of 55 in 1893. The British Raj now exerted supremacy over the whole Indian sub-continent, bringing together a nation of such cultural diversity that it is unparalleled in the world.

The rise of Sikh militancy, particularly in the years following 1981, was due to the rapid progress the Punjab was making in food production, industry and other aspects. Vast numbers of Sikhs had already established themselves in Western countries such as Canada, Britain and the United States, with smaller pockets in countries like Germany. The settlements abroad also brought unexpected benefits for those left behind in the Punjab. Sikhs living in their new countries had done tremendously well because of hard work and an enterprising spirit. These settlers abroad became one of the largest sources of foreign exchange for India. Many Sikhs living in the West, for example, still own massive tracts of land in their homeland. The Punjab, where Sikhs form a slight majority over the Hindu Punjabis, has prospered rapidly particularly because of the miraculous green revolution which made India not only self-sufficient in food production for the first time in many years, but also an exporter of food. This new-found prosperity brought fresh demands for greater attention to the Punjab by the central government in New Delhi.

In September 1981, the Akali Dal, the mainstream political party in the Punjab, forwarded a set of 45 demands to the government of Prime Minister Indira Gandhi. Some of the demands had their origin in what is known as the Anandpur Sahib resolution, adopted in 1973, which had asked for the merger of several Punjabi-speaking areas in the next-door Haryana State with the Punjab to form one administrative state. But the resolution also had another element: namely, greater freedom from central control. The situation was not too different from what Canada faces in the tug between Ottawa and the provinces.

The resolution read: 'In this new Punjab [formed by the merger with neighbouring territories], the Central intervention should be restricted to defense, foreign affairs, posts and telegraphs, currency and railways.'

It was just the beginning of the tug of war between New Delhi and the Akali Dal. New Delhi now faced a major problem: If it gave in to these demands from the Punjab, how would it stop other States of India from demanding similar rights?

The set of 45 demands handed to New Delhi in 1981 asked for sweeping reform in several areas. They fell generally into three categories: religious, economic, and relations between the central government and the Punjab administration. The chief religious demands were that Amritsar, the seat of the Holy Shrine of the Golden Temple, be granted Holy City status, which would result in banning the sale of alcohol, tobacco and meat within the walled city, and that a radio transmitter be installed at the temple to relay *kirtan* (the Sikh name for prayer) outside of State control. Radio stations in India are now State-owned. Further, the Sikhs wanted to be able to wear the *kirpan* or ceremonial dagger on domestic and international airline flights, and they wanted a new law to bring all historical Sikh temples in the entire country under one administration.

Quite clearly, some of the demands could be met easily, but others were hard to swallow for New Delhi. It was pointed out, for example, that declaring a city a Holy City would not be in keeping with the secular policies of the nation. However, the government would agree to ban the sale of offending items in an area close to the Golden Temple and a sacred Hindu Temple. The Indian constitution allows Sikhs to wear the ceremonial dagger, but carrying the *kirpan* was banned on domestic flights after a hijacking in 1981. It was explained that international airline regulations would not permit the carrying of a dagger. Another demand, also of a religious nature, was the centralized control of all Sikh temples under an umbrella organization. There was fierce resistance from

Sikhs living outside the State to such centralization of authority, said Mrs Gandhi's government.

Further disputes arose between the central government and the Akali Dal on the basis of the distribution of river water, demands for which were increasing as the flourishing agricultural industry grew. In the 1970s Mrs Gandhi had promised the transfer of the city of Chandighar to the Punjab, so long as agreement could be reached to transfer some Hindi-speaking areas in the Punjab to Haryana.

The hardest thing to negotiate, said Mrs Gandhi's government, was the demand for watering-down of central government control of the provinces, which she said went against the grain of national unity.

As negotiations stumbled and the Akali Dal began a series of agitation movements, including stoppage of work, stoppage of railways and demonstrations aimed at clogging the jails and courts, another element was creeping in to exploit the situation. Among the chief proponents of a hard-line approach was the All India Sikh Students' Federation and a man called Sant Jarnail Singh Bhindranwale, who was at first courted by New Delhi, until he became an uncontrollable rebel.

Fuelling those elements of lawlessness that would overtake the Punjab in the coming years were repeated killings generated by a sectarian feud between fundamentalist Sikhs and a sect called Nirankaris, an offshoot of Sikhism who believe in a formless God who can only be realized by the presence of a living Guru. Sikhs felt that the New Delhi government was actively supporting the Nirankaris to deliberately erode the Sikh faith as it had been preached since the days of the first Guru. Various clashes occurred in 1978 between fundamentalist Sikhs and the Nirankaris which finally climaxed with the killing of Gurbachan Singh, spiritual leader of the Nirankaris, in 1980.

It was Sant Jarnail Singh Bhindranwale who virtually began an undeclared war with New Delhi and subsequently plunged the Punjab into a vicious circle of violence that would shake India and pose a threat to its national security worse than at any other time since its birth and the partition with

Pakistan. The saint-soldier, who would later turn the Golden Temple into a fortress, made proclamations to his small but loyal band of followers that openly incited communal violence between Sikhs and Hindus in the prosperous State of Punjab. Bands of his followers would roam around the province, shouting slogans like 'Khalistan Zindabad', meaning Long Live the State of Khalistan, a name coined by rebel Jagjit Singh Chauhan who now lives in England. At about the same time, a newspaper editor who had criticized the killings of Nirankaris was himself murdered.

Bhindranwale was arrested by police in connection with the murder of Lala Jagat Narain, prompting massive violence in the streets of the Punjab, which in turn prompted the police to fire on a crowd of people. The scene had been set for a fire in the Punjab which would be fuelled from both within and outside India by small pockets of militants living in Western countries. But at this time the vast majority of Sikhs living overseas remained unaffected by the turmoil that was embroiling their home province.

2

Flames of Hatred

The Punjab was overtaken by serious violence beginning in
early 1981. As the Akali Dal continued its series of agitations
for greater religious and economic independence, Bhindranwale and the All India Sikh Students' Federation were
consolidating their position, establishing a solid and loyal
group of supporters who were willing to lay down their lives.
The men surrounding Bhindranwale had already decided to
take the advice given by Mahatma Gandhi in 1931.
Bhindranwale advocated the use of the sword, openly and
defiantly.

When Indian authorities released him from jail after
dropping the earlier charge of murder, he promptly moved
his headquarters within the boundary of the Golden Temple.
At about the same time, roving gangs of extremists riding
motorbikes carried out widespread killings in the Punjab,
while nervous police officers often resorted to firing on
crowds of protesters when things got out of control.

The vicious circle of brutality had begun. In May of 1982,
India declared both a militant group called Dal Khalsa and
Dr Chauhan's National Council of Khalistan unlawful
organizations. Shortly thereafter, Amrik Singh, president of
the Sikh Students' Federation, was taken into custody and
this prompted another wave of violence. When things finally
began to go out of control with New Delhi still unable to
satisfy the demands of the moderates, direct presidential rule

was imposed in 1983. The Punjab government resigned because of the scale of violence that had gripped the province. There were virtually daily incidents of shootings, despite deployment of massive contingents of the army and police in the State.

Bhindranwale had now moved into the sacred Akal Takth within the Golden Temple and directed his operations from the holy site. Things were coming to a head in the Punjab now. In January 1984, the Akali Dal, already swamped and taken over by extremist groups, decided to burn copies of the Indian constitution. By now Hindu extremist organizations such as the Hindu Surakhsha Sumiti, in essence a Hindu defence committee, in turn began sacreligious actions against the Sikh religion. In one instance the militant Hindu unit damaged a model of the Golden Temple and defiantly burned a picture of a Sikh Guru. That led to a divisive conflict between Hindus and Sikhs, who had for generations lived side by side in the Punjab.

The gravest crisis would follow shortly after 2 June 1984, when New Delhi sent in the troops armed with plans for Operation Blue Star. A day later, the entry of foreigners into the Punjab was banned and the border with Pakistan sealed up. The Siege of the Punjab had begun. A 36-hour curfew was announced as troops moved into position around the Golden Temple, where Bhindranwale had already made preparations for battle. Fortifications were built around several areas within the temple, with sandbagged battle stations manned with machine guns and an assortment of other weapons. The Indian Army would find that incredible preparations had been made, including effective military-type communication between Bhindranwale's various forces deployed within the temple. The stage was now set for turning the most sacred shrine of the Sikhs into a battlefield.

On the evening of 5 June, the army addressed the militants within the temple to surrender, but only 129 came out with their arms raised. By nightfall Operation Blue Star was in full motion and the temple became a slaughterhouse, for both army units and the Sikhs holed up in it. In the first few hours of the operation, as a battle raged for control of the periphery of the temple area, moderate Akali Dal boss Harchand Singh Longowal was arrested by the army with over 300 others. But Bhindranwale was determined to stand his ground.

Early on the morning of 6 June, as the world watched in awe, the army moved in heavy armour, including an armoured personnel carrier. But this was immobilized by anti-tank rockets fired from within the Akal Takth fortification. Then the army moved into the Akal Takth itself, using the searchlights of a tank to blind militant gun positions. Even then Bhindranwale would not give up. There was room-to-room fighting and by morning the superior training and armour of the Indian Army had overcome major resistance.

Mopping-up operations continued the next day. Bhindranwale's bullet-riddled body was discovered in the basement along with that of Amrik Singh, president of the Sikh Students' Federation. Also killed in the action was former Indian Army General Subheg Singh, who had teamed up with Bhindranwale. The main seat of the Sikh religion, the Akal Takth, was badly damaged. Simultaneously with the operation against the Golden Temple, other raids were launched against other Sikh temples.

The casualties were enormous. The army itself had suffered almost 80 dead and over 250 wounded. The casualties on the side of the men defending the temple were much, much higher. Official government figures indicate a death toll of 500 with 86 injured. But Sikhs who witnessed the assault swear that the loss of lives was in the thousands. Hundreds of militants were taken into custody and arms and ammunition of all descriptions were seized. Inside, the army also found a crude hand-grenade factory.

Mrs Gandhi had won the battle but had lost the hearts and souls of her Sikh countrymen. Sikhs world-wide were reduced to tears by the attack on the Golden Temple. The soul of the Sikhs had been violated. The question was being asked in every Sikh household: Was it really necessary to send the army in? Was it necessary to drag out the negotiations on Sikh demands until they reached such a tragic climax?

The wounds created by the attack on the Golden Temple

would take a long time to heal, but just when it looked like the brave saint-soldiers were beginning to forgive, another tragic event occurred. Two Sikhs, acting in isolation, had made up their minds to avenge the attack on the Golden Temple.

Early on the morning of 31 October 1984, Mrs Gandhi's trusted Sikh bodyguards opened fire on the Prime Minister after first saluting her as she walked down the path towards her office. The most powerful woman on earth, who had vowed to die rather than see her country split up, fell to the ground, her frail body jerking violently as she was riddled with dozens of bullets from machine guns. Like Mahatma Gandhi, she too had fallen prey to religious fanatics. The one who killed the Mahatma was a Hindu. He did it because Gandhi was giving away too much to the Moslems. Mrs Gandhi was cut down because she wanted to give too little to the Sikhs.

Her departure left an immense power vacuum in the country, and the moment the country recovered from the initial daze over her assassination, there followed three days of hell for Sikhs living in the Indian capital, New Delhi. There is no excuse for what happened for three days after her death. As the reigns of power fell into the hands of the young, charismatic Rajiv Gandhi, angry mobs bent on vengeance over his mother's death carried out a systematic massacre of innocent Sikhs, sometimes right in front of police. Sikhs were burned alive, their women raped, and their homes ransacked.

Granted, the country was in a state of profound shock during the days that followed Mrs Gandhi's death and her heir was torn between grief and taking his first steps towards exercising power. But there was no reason for the lack of police intervention in many spots where the atrocities were being committed against Sikhs who had played no role in the assassination. The mobs had made the mistake that is so easy to make. The mistake of generalization. A Sikh is a Sikh. Therefore, because her assassins were Sikhs, the rest of the Sikhs were just as guilty. The bloodshed would not stop until the army moved in and peace groups, formed by Hindus and Sikhs grieved by the scenes of cruelty, marched on the streets of New Delhi chanting 'Hindu-Sikhs Bhai Bhai', meaning

Sikhs and Hindus are brothers. Again, it became clear that the lives of many Sikhs were spared because Hindu families, at great risk to themselves, gave Sikhs shelter in their homes as fellow countrymen.

Such are the entrapments of religious strife. In Ireland, in Lebanon, between the Israelis and the Arabs and in India. Each time the vicious circle goes on and on, and the ones who pay a price are usually the innocent who want no part of the fanatical ideologies of the few. But they are the few who hold the guns and billy clubs and iron bars. Intimidation works wonders. And that's the trick of the terrorist trade.

As Rajiv Gandhi became the power-broker in New Delhi he recognized that the situation in the Punjab was the greatest challenge India has had to face in modern times. He knew the time was running out to stop the flames of secession. The new Prime Minister is a man of modern times, not having the handicaps of the old school of thought that had plagued successive administrations in dealing with legitimate Sikh demands that stopped short of autonomy. One of the first things he did was to appoint a commission of inquiry to look into the riots that had ripped through New Delhi following the assassination of his mother.

He also moved with amazing speed to begin negotiations with moderate leader Harchand Singh Longowal to defuse the crisis that had created so much bloodshed in his country. On 24 July 1985 he reached a wide-ranging accord with Longowal, answering many of the demands that had been made for years by Sikhs.

The accord offered compensation to the families of all innocent victims of the Punjab turmoil since 1982, abandoned a previously enforced quota on the number of Sikhs who could enrol in the armed forces and made merit the only criteria for selection, extended the inquiry into riots following Mrs Gandhi's death to other areas besides Delhi and offered to rehabilitate those Sikh soldiers who had defected from the army in protest against the raid on the Golden Temple. Further, Gandhi promised to consider legislation to put in place an act of parliament to form a unitary body to manage temples throughout India.

Another deal he offered with the olive branch was the transfer of Chandighar to the Punjab and other territorial issues to a commission. Longowal in turn agreed to abandon any talk of a separate Sikh State and, further, that all considerations of more regional power would only be given when they did not clash with the need for the unity of the country. Agreement was also reached on some of the economic disputes concerning river waters.

Gandhi ordered the end of direct rule for the Punjab and called State elections. The Congress party was handed a stunning defeat, but democracy had won. An unprecedented 60 per cent of the voters would turn out to send a message to those who were preaching turmoil.

Sadly, though, as the Akali Dal boss Surjit Barnala romped to victory, Longowal was not present to see it. He was gunned down on 20 August by militants who called him a traitor to the Sikh cause.

Overseas, when Longowal died, there was little show of grief among hardliners. But at least they did not perform a pitiful dance of death as some had done in London and Vancouver when Mrs Gandhi died.

3

Militants outside India

A massive majority of Sikhs living abroad in countries such as
Canada, Britain and the USA had remained largely
unaffected by the feud between New Delhi and the Punjab
extremists. But when the Golden Temple was raided, a
sudden wave of emotion overtook thousands upon thousands
of overseas Sikhs. In Vancouver, for example, 25,000 Sikhs –
a majority of them moderates – marched on the streets to
show their disapproval over the raid on their Vatican. Their
grief was understandable. They had been shocked, utterly,
that an army would be needed to flush out a group of
extremists from the seat of Sikhism's soul, the Golden
Temple. It was around this time that small militant groups
operating overseas gained momentum and membership
amongst Sikhs who had previously ignored the hue and cry in
the Punjab.

It is still true, though, that membership of militant groups
involves only a small minority of Sikhs who have settled in
Canada, Britain and the USA. Many have been in the West
for so long that they do not even care about what's happening
in a country with which they have severed their former ties.
Sikhs living in the West among other citizens of those
countries are enterprising people who have built a solid base
for their families. They are among the best-housed and
employed people in their new countries.

But numerous Sikh separatist groups now operate within

Canada, Britain and the USA, and monitoring their activitie is costing counter-terrorist agencies a bundle of money. Fo example, in Canada, Sikh militants are now one of the twe major areas in which the manpower of the Canadian Security Intelligence Service is deployed. Incidents of beatings and intimidation of moderates – and there have been many – are not a concern of intelligence agencies. Those aspects are deal with by regular police forces. However, another aspect of Sikh militancy *is* of major concern.

This is the use of the host countries as launching pads for terrorist activities in India.

Most militants interviewed by intelligence agencies in the three Western nations always insist that they will abide by the laws of the country in which they live – that crimes will not be committed within those countries. However, to the Western nations, as Margaret Thatcher, Prime Minister Brian Mulroney and the Reagan administration have made clear, the practice of using them as safe havens from which to launch strikes against India is not acceptable either.

It is not uncommon for extremist groups to preach that no nation has been born on the conference table. It is not uncommon for various splinter groups to hold press conferences and claim responsibility for assassinations in India. The tragic downing of Air India 182 was among several incidents that have raised supreme concern within Western intelligence agencies.

For example, in a top-secret FBI operation code-named Operation Rite Cross, an FBI agent infiltrated a small radical group of Sikhs belonging to an obscure cell called Black June. The four men had approached Frank Camper, who ran a school called 'The Mercenary School' in Dolomite, Alabama. Camper told the FBI later that the four wanted training in the manufacture of time bombs, attacking armoured vehicles, assassination techniques and even how to blow up trains.

Camper said the men made it clear that their aim was to engage in sabotage in India. But that's what they said to Camper. The FBI knew otherwise. The four were arrested later for plotting the murder of an Indian minister, and one of them was also indicted for plotting to assassinate Prime

Minister Rajiv Gandhi during a visit to Washington in the summer of 1985. Two of the wanted men, Lal Singh and Ammand Singh, escaped the FBI dragnet.

That is why FBI agents visited Vancouver, according to intelligence authorities in the United States. It was the FBI's first brush with Sikh militancy. They wanted to be briefed by the CSIS, who were by now the acknowledged experts in this field. Subsequent to Operation Rite Cross, though, the FBI realized the hitherto unsuspected existence of a whole network of militant groups based in various cities in the United States. They now monitor those groups' activities on a constant basis.

Early last year, Canadian authorities and the Scotland Yard anti-terrorist squad intercepted two men carrying an Israeli-made UZI machine gun. The gun had been disassembled and its parts were being carried in two separate bags. It has never become clear why the deadly weapon was being ferried to London from Vancouver, although one of the men proclaimed later that he was ordered under threat to carry the gun – known as the perfect weapon for an assassination because of its lethal power and concealability – to a high official of the Khalistan movement in London.

It is not uncommon nowadays for the British Security Service to ask England-based militants about their connections with their counterparts in Canada. The BSS knows about considerable cross-traffic and visits between members of some groups in Canada and those based in England. And late last year, during a visit of Prime Minister Rajiv Gandhi, who must today be one of the most targeted leaders in the world, British authorities foiled an alleged plot to assassinate him by four men resident in the Birmingham area. The case is now before the courts. Their guilt or innocence will be proven there.

In London, Khalistan Liberation Movement chief Dr Jagjit Singh Chauhan continues preaching his cause from a building he has named Khalistan House. The former Punjab state minister often shows off letters he received from Sant Jarnail Singh Bhindranwale, who died in the storming of the Golden Temple. Bhindranwale had urged Chauhan to carry on the struggle from outside the country to bolster the movement he was spearheading. Chauhan has good contacts among various groups in Canada, the USA and Germany, but his overseas travel has been curtailed since he has been barred from entering Canada and the USA. His Indian passport also has been revoked, and India raised a storm when he was allowed to enter the United States on that cancelled passport. The British government has also instructed Mr Chauhan to curtail his activities and confine them within the bounds of democracy and the laws of the land.

Among the chief groups operating in Canada, with strong ties both in the UK and the USA, is the World Sikh Organization. This is a mainstream body with representation in the thousands in the three countries where Sikhs live in massive numbers. The body is led by prominent California multimillionaire Didar Singh Bains, who travels back and forth between cities where he has a following in a private jet appropriately marked 'Khalistan 1'. It surely must be the envy of the US president, who is flown around in 'Air Force 1'.

The group led by Bains includes fairly large numbers of intellectuals, among them former Indian Army General Jaswant Singh Bhullar, who fled charges of sedition in India and escaped shortly before the Temple was raided. Bhullar said he left because Bhindranwale wanted him to leave. The renegade general had been instrumental in the taking of the capital of East Pakistan when India actively supported the guerrillas of the Mukti Bahini, who wanted to separate from West Pakistan. The war that followed led to the creation of Bangladesh. The stated policy of the WSO is the creation of a separate Sikh state. The organization says it plans to do that by lobbying rather than violence. However, in many cases where Sikhs have faced charges resulting from alleged criminal offences the WSO has rushed to offer financial aid for the defence of those accused. That is their right, of course, so they are within legal bounds.

The other major organization, much more militant than the WSO, and perhaps also the organization that has the capability to become the most powerful, is the International Sikh Youth Federation. Intelligence agencies in the West say the group is still in the organizational stage, but it is one of the primary targets of surveillance by security authorities. The group is governed by a nine-member High Command which is spread between Canada, England, the USA and India.

The ISYF regards other less militant groups as paper tigers and has been actively taking control of major Sikh temples in Western countries in order to obtain a stage for propagating its cause. The branch in Canada is headed by a Vancouver-area resident by the name of Satinderpal Gill, who makes frequent trips to Pakistan, considered by many radicals to be a natural ally of the struggle to carve out Khalistan. Canadian intelligence sources know the ISYF pumps enormous amounts of money into Pakistan and India, but little is known about what happens to it.

The Canadian branch of the ISYF has been bolstered in recent days by the arrival of Lakhbir Singh Brar, a nephew of Sant. Jarnail Singh Bhindranwale. Brar is called a 'convenor' of the ISYF. He came to Canada in April 1985 from Abu Dhabi where he was a well-to-do businessman. Brar fled Abu Dhabi after his brother, Jasbir, was shuttled back and forth between England and the Philippines and finally handed over to the Indians. Brar has applied for refugee status in Canada.

Another new arrival is former England resident Harpal Singh Ghumman, real name Harjinderpal Singh Nagra, one of the founders of the ISYF branch in England. Ghumman, a hardline fundamentalist, has a degree in law and is one of the masterminds of the movement. Early in 1985, Ghumman married a Canadian resident and applied to stay in Vancouver. That brought another of his friends here too. His name is Pushpinder Singh, real name Mohaninder Singh Sachdeva, a man with a PhD degree who is extremely bright. Pushpinder Singh, after leaving the Philippines where he lived, entered Canada by way of Mexico and then slipped into British Columbia where he has now applied for refugee status on the basis that if he's sent back, he'll face persecution in India.

The third group of some significance is the extreme fundamentalist group called Babar Khalsa, first formed in India in 1978. The overseas unit is headed by Vancouver area

Sikh Talwinder Singh Parmar, also known as Harpav Singh Parmar, who has often made his former aide Surjan Singh Gill pronounce him the leader of 14 million Sikhs. Gill broke away from his group in April 1985 after a personal row. Parmar's following is restricted to about 400 in Canada, mostly in Ontario, with some support in England and Germany, intelligence analysts say. Parmar, who calls himself a *Jathedar*, a commander who fights injustice, lives in a plush house and owns three cars. Authorities believe he's financed by his followers and a rich Vancouver area businessman. Babar Khalsa has often offered support to hijackers of Indian Airlines jets on trial in Pakistan, saying it was a symbolic gesture of defiance and did not cause a loss of life.

A publication put out by the Babar Khalsa, called 'The Case for Khalistan', states: 'The symbolic hijacking of an Indian airliner by Sikh youth is simply a message to India, and the world, that now the Sikhs mean business.'

But the Babar Khalsa could be advised, and rightly, by millions of Sikhs around the world, to speak for itself. Sikhs as a community do not endorse hijackings, symbolic or otherwise.

The terrorist attack on Air India prompted the government of Canada to rethink its strategy for dealing with militant groups. External Affairs Minister Joe Clark, during a visit to India, offered wide-ranging collaboration between the intelligence authorities of the two countries and an extradition treaty to appease the Indians. Clearly, the terrorist attack was an embarrassment to the Canadian government in view of previous warnings from India. Further, Ottawa also paved the way to deporting Sikhs whose applications for refugee status had been denied, thus lifting a moratorium that had been in effect since 1984. But Clark rightly pointed out that it would be unfair to label all Sikhs militants because of the actions of a few.

It would be easy to fall into the trap of generalization. The only reason those individuals who carried out the attack against the Air India jet are identified as Sikhs is because they themselves define themselves as members of Sikh militant groups. Further, the identification is relevant because the attack was carried out in the misguided belief that all is justified in the name of religion. But the public, in India, England, Canada and elsewhere, would be making a tragic error if it misconstrued the use of this label and concluded that all Sikhs are terrorists. Not so.

That was the error made by those unruly and mindless mobs who massacred Sikhs simply because they were Sikhs following the assassination of Indira Gandhi. No, the true followers of the noble Khalsa deserve better than that. It would be the equivalent of construing the actions of the Irish Republican Army to represent the tenets of the Christian faith, or for that matter those crusaders of the past. Or of construing the actions of Shia militants in the Middle East as the teachings of the faith of Islam.

No religion preaches violence. But a handful of people in the Christian, the Hindu, the Moslem and the Sikh faith, in their own twisted logic, believe that God has given them the licence to inflict brutality on innocent people. It is a wonder the Lord above hasn't spoken out. It's time He said: 'For God's sake, don't do it in my name.'

The downing of the jet was a joint operation between two very small groups that can claim only very limited support in Canada. Furthermore, the RCMP believe, the operation was carried out by a cell-structure of no more than 12 people. Three or four men organized the scheme. Another made the bombs, somebody else booked the tickets, then another person picked them up. Two different men then delivered the deadly cargo to Vancouver International Airport.

The act caused as much grief within the Sikh community as outside. Most Sikhs were so stunned that they refused to believe that any member of the Sikh faith, which preaches love and tolerance, could do a thing like that to innocent, defenceless civilians. That is why, when the airliner went down, the main Sikh temple in Vancouver initiated 15 days of mourning for all the dead, not just the Sikhs.

APPENDICES

1. The Victims of Air India Disaster

CREW:

1. CAPT. HANSE SINGH NARENDRA
2. COPILOT SATWINDER SINGH BHINDER
3. FLT. ENGINEER DARA DUMASIA
4. SAMPATH LAZAR
5. RIMA BHASIN
6. JAMSHED DINSHAW
7. PAMELA DINSHAW
8. SHYAMA GAONKAR
9. SANGEETA GHATGE
10. LEENA KAJ
11. NELI KASHIPRI
12. SHARON LASRADO
13. RITA PHANSEKAR
14. SUSEELA RGHAVAN
15. ELAINE RODERICKS
16. BIMAL SAHA
17. KARAN SETH
18. SUNIL SHUKLA
19. SURENDRA SINGH
20. INDER THAKUR
21. KANAYA THAKUR
22. NOSHIR VAID

OTHER STAFF OF AIR INDIA:

23 SHYLA AURORA
24 FREDDY BALSARA
25 CHAND BHAT
26 LEENA BISEN
27 SUMAN KHERA
28 MARAZBAN PATEL
29 IRENE SHUKLA
30 ANNE TRAVASSO

PASSENGERS:

31 AGGARWAL RAHUL
32 AHMED INDRA
33 AHMED SARAH
34 ALEXANDER A. DR
35 ALEXANDER SIMON
36 ALEXANDER
37 SINGH USHA
38 SINGH AKHAND DR
39 ALLARD GOLLETTE
40 ANANTRAMAN
41 ANANTRAMAN
42 ANANTRAMAN BHAWAN
43 ANTHONI DESA
44 ASHIRVATHAM RUTH DR
45 ASHIRVATHAM A.
46 ASHIRVATHAM SUNITA MISS
47 BALAMARAN S. MRS
48 BALAMARAN N.
49 BALASUBRAMAN R.
50 BAJAJ ANJU
51 BEAUCHESNE G.
52 BEDI V.
53 BEDI JATIN MASTER
54 BEDI ANU
55 BERAR J.
56 BERRY SHARAD MR

57 BEDI SARAJ
58 BERAR J.
59 BERAR
60 BHAGAT ADARSH
61 BHAGWANTI B.
62 BHALLA NIRMAL
63 BHALLA DALIP MASTER
64 BHALLA MANJU
65 BHARDWAJ HARISH
66 BHAT TINA MISS
67 BHAT BINA MISS
68 BHAT VINU MR
69 BHAT P. V.
70 BHAT C. P.
71 BHAT S.
72 BHATT DEEPAK MASTER
73 CASTONGUAY R.
74 CASTONGUAY R.
75 CHANDRASEKHA S.
76 CHATLANI NEETAM
77 CHATLANI MARC
78 CHATLANI MALAM
79 CHEEMA S.
80 CHOPRA S. R. MRS
81 CHOPRA J. R. MR
82 DANIEL VARGHESE MR
83 DANIEL RUBY MISS
84 DANIEL ROBYN MISS
85 DANIEL CELINE MRS
86 DAS ARUPT
87 DAS RUBY
88 DESA A.
89 DESOUZA R.
90 DHUNNA R.
91 DHUNNA SASHI
92 DHUNNA SUNEAL MASTER
93 DHUNNA BHAG
94 ENAYATI A.
95 FURDOONJI HOMAI

96 GADKAR ANITA
97 GADKAR
98 GAMBHIR J.
99 GAMBHIR A.
100 GAMBHIR S.
101 GOCNE RITU MISS
102 GOPALAN KRISHN
103 GOSSAIN ARUN MASTER
104 GOSSAIN APAPARNA MISS
105 GOSSAIN KALPANA
106 GREWAL DALJIT
107 GUPTA SANTOSH
108 GUPTA SASHI
109 GUPTA A. MISS
110 GUPTA G. MASTER
111 GUPTA RAJESH
112 GUPTA B.
113 GUPTA S.
114 GUPTA A. MASTER
115 GUPTA RAMVAT
116 GUPTA ANUMIT
117 GUPTA S.
118 HARPALANI R. MRS
119 HARPALANI D. MISS
120 HARPALANI S. MISS
121 JACOB P. MR
122 JACOB J. MISS
123 JACOB J. MASTER
124 JACOB A. MRS
125 JACOB J. MISS
126 JAIN ANOOPUMA
127 JAIN INDU
128 JAIN RUCHI
129 JAIN PRAKASH
130 JAIPURIA MALA
131 JALAN DEV
132 JALAN ANITA MISS
133 JALAN VINAY (INFANT)
134 JAMES JOSEPHINE SR

135 JETHVA V.
136 JETHVA Z.
137 JOB ALEYKUTTY MRS
138 JOB TEENA MISS
139 JUTRAS RITA
140 KACHROO M.
141 KALSI INDRA
142 KAMMILA RAMA
143 KAPOOR SANTOSH
144 KAPOOR SHARMILA
145 KAPOOR SABRINA
146 KASHIPRI ATHIKO
147 KAUR SUKHWINDER
148 KAUR PARAMJIT
149 KAUR GURMIT
150 KAUSHAL BISHAN
151 KELLY BARSA
152 KHANDELWAL C.
153 KHANDELWAL M.
154 KHERA RASHI
155 KOCHNER S.
156 KUMAR R. MR
157 KUMAR C. MRS
158 KUMAR MANJU
159 LAKSHMAN P.
160 LAKSHMAN K.
161 LAURENCE SHYAMALA MISS
162 LAURENCE NICOLA MISS
163 LAZAR S.
164 LAZAR S.
165 LEGER G.
166 LOUGHEED D.
167 LULLA MUNISH MR
168 MADON SAM
169 MAINGUY L.
170 MALHOTRA ATUL
171 MAMAK R.
172 MANJANI N.
173 MARJARA DEVENDRA

174 MARJARA SEEMANT
175 MARTEL A.
176 MEGHNA SABHARWAL
177 MEHTA KISHORE MR
178 MEHTA CHANDRA MRS
179 MEHTA NEESHA MISS
180 MEHTA NILISH MR
181 MERCHANT NATASHA
182 MINHAS BALIHNDER MRS
183 MINHAS KULBIR MISS
184 MOLAKALA PRABHAVATI MRS
185 MUKERJI NISHIT
186 MUKHI RENU
187 MULLICK DEEPAK
188 MURTHY
189 MURTHY
190 MURTHY
191 MURTY
192 MURUGAN
193 MURUGAN
194 MURUGAN S.
195 MURUGAN G.
196 NADKARNI DEVEN MR
197 NADKARNI RAHUL MR
198 NAYUDAMMA YELAVARTHY DR
199 PADA VISHNU MR
200 PADA ARTI MISS
201 PADA BRINDA MISS
202 PALIWAL MUKUL MASTER
203 PATEL BIPIN MR
204 PATEL B.
205 PATEL MOHAN
206 PRYIA THAKUR
207 PURI VEENA
208 PURI AMIT MASTER
209 QUADRI SYED MR
210 QUADRI RUBINA
211 QUADRI ARSHYA
212 QUADRI A (INFANT)

213 RADHAKRISHNA NAGU MR
214 RADHAKRISHNA JYOTHI MISS
215 RADHAKRISHNA THEJU MASTER
216 RAGHUVERRAN RAJIV MR
217 RAGHUVERRAN VASANTHA MRS
218 RAI KIRAJIT MISS
219 RAMACHANDRAN PRATIBHA MISS
220 RAMASWAMI JANAKI MISS
221 RAMATHULLA MOHAMMED MR
222 RAUTHAN B.
223 RAUTHAN POOJA
224 SADIQ S. MRS
225 SAGI SUJATA
226 SAGI KALPANA MISS
227 SAGI KAVITA MISS
228 SAHU P.
229 SAHU P.
230 SAHU R.
231 SAKHAWALKAR DATTAR MR
232 SAKHAWALKAR S. MISS
233 SAKHAWALKAR USHA MRS
234 SAKHAWALKAR S. MR
235 SAKHAWALKAR S. MR
236 SANKURATHRI S. MR
237 SANKURATHRI S. MISS
238 SANKURATHRI M. MRS
239 SARANGI RAJSRI MISS
240 SETH KALPANA
241 SETH SHILPA
242 SETH SADHANA
243 SETH SATISH
244 SETH (INFANT)
245 SHAWNEY O. P.
246 SHARMA SHAKUNTLA MRS
247 SHARMA UMA MRS
248 SHARMA S. MISS
249 SHARMA OMPRAKASH MR
250 SHARMA MANMOH MR
251 SHARMA N. MASTER

252 SHARMA S. MR
253 SHARMA A. MASTER
254 SHARMA U. MASTER
255 SHARMA R. MISS
256 SHARMA SUMITRA
257 SHARMA INDU MRS
258 SHARMA V. MISS
259 SHARMA N. MR
260 SHARMA SUNDEE
261 SHARMA SHARVAN
262 SHARMA SHYAM
263 SINGH JAGJIT
264 SINGH ABHINAV
265 SINGH MUKHTIAR
266 SINGH ANAND
267 SINGH SUBHANA
268 SINGH SHALINI
269 SINGH SHALINI
270 SINGH BALBIR
271 SINGH J. MISS
272 SINGH R. (INFANT)
273 SINGH AJAI MASTER
274 SINGH J.
275 SINGH AMAR MASTER
276 SINGH DARA MR
277 SINGH R. MISS
278 SINHA R. K. PROF.
279 SONI RITU MISS
280 SONI PANKAJ MASTER
281 SONI MONICA MISS
282 SONI U. MRS
283 SRAN PRIMALJIT MISS
284 SRIVASTAVA
285 SUBRAMANIAN L. MRS
286 SUBRAMANIAN G. K. MR
287 SUBHRAMANIAN
288 SUBHRAMANIAN K. MR
289 SUBHRAMANIAN J. MRS
290 SUBHRAMANIAN U. MISS

291 SUBHRAMANIAN S. MISS
292 SUTRAS R. MISS
293 SWAMINATHAN A.
294 SWAMINATHAN I.
295 SWAMINATHAN P. MISS
296 THACHETTU IVY MISS
297 THAKUR VISHAL
298 THAMPI VIJAY
299 THOMAS K. K.
300 THOMAS
301 THOMAS
302 THOMAS
303 TRAVASSO L. M.
304 TRAVASSO C. A.
305 TRAVASSO L. L. MISS
306 TRIVEDI NIRMAL
307 TRIVEDI
308 TRIVEDI
309 TUMKUR CHITRA MISS
310 TUMKUR RAMOH MR
311 TURLAPATI SANJAY
312 TURLAPATI 11-YEAR
313 TURLAPATI F.
314 TURLAPATI DEEPAK MASTER
315 UPPAL PARMINDER
316 UPPAL KULDIP
317 UPRETI HEMA
318 UPRETI VIKRAM
319 UPRETI G. C. DR
320 VAZ JULIET
321 VENKATESAN GEETHA
322 VENKATESHWARAN TRICHU MR
323 VERMA BALVINDER
324 VISHAY THAKUR
325 WADHWA SERINA MISS
326 WADHWA AKHIL MASTER
327 YALLAPRAGADA MURTHY MR
328 QUADRI SHAISTA MRS
329 JALAN SHILA MRS

VICTIMS OF NARITA BOMB:

DEAD:

1   KODA IDEHARU
2   ASANO HIDEO

INJURED BUT LUCKY TO BE ALIVE:

1   ASAKURA MASAHARU
2   ASANO TESTUYA
3   YOSHOKA MASAHIRO
4   SOGA TOSHINOBU

2

The Tape Transcript of the Cockpit Voice Recorder from 0643:42 on AI-182

ATC Timings (Z) (1)	Channel Identi- fication (2)	Voice Identi- fication (3)	Text (4)
0643:42	A, C	AI-182	And Air India 182 position
0643:48	A, C	Shanwick	—
0643:56	A, C	AI-182	Air India 182 Shanwick position is 51 N 20 W at 0643 level 310 Estimate 51 N 15 W at 0704.
0644:15	A, C	Shanwick	Roger select position after 15 minutes please
0644:20	A, C	AI-182	182
0646:38	A, C	AI-182(B)	Shanwick AI-182
0646:41	A, C	Shanwick	AI-182 Shanwick could you say your next position after 51 N 15 W please
0646:48	A, C	AI-182(B)	After that we have 51 N and 8 W, 51 N 8 W and thereafter Bunty, Bunty.
0647:02	A, C	Shanwick	Roger AI-182 Shanwick confirm 51 N 20 W 0643 310, 51 N 15 W 0704. Next 51 N 8 W. Go ahead
0647:14	A, C	AI-182(B)	That is affirmative
0647:16	A, C	Shanwick	Roger at 15 W Call Shannon 135.6
0647:21	A, C	AI-182(B)	135.6 will do Sir

0647:22	A, C	Shanwick	Roger Good day
0647:23	A, C	SR-101	Shanwick Swiss Air 101 position over
0647:26	A, C	Shanwick	101 Shanwick standby one
0647:27	A, C	SR-101	Roger
0650:15	A	—	Cough
0655:19	A		Cough cough cough
0659:41	A, C	1831	Shannon 1831
0659:45	A, C	1831	Is flight level 330 available?
0659:55	A, C	—	Shannon, flight level 330 available
0700:05	A, C	—	— — — — FL 330
0700:13	A		A couple of forms
0700:16	A		— — of Air India
0700:17	A		What about those forms
0700:18	A		Haaji [yes]
0700:21	A		Aur wo custom ka form [those custom forms]
	A		— —
0700:25	A		Baad main — — [later]
0700:29	A	B	Dinshaw
0700:29	A	JD	Yes Sir
0700:30	A	B	Do me a small favour
0700:31	A	JD	What's that
0700:32	A	B	Ekdam end pe, 54 seat pe [right at the end – on seat 54] a boy is sitting there. Inder Thakur knows. He just wanted to have a look in the Cockpit
0700:43	A		Where is he
0700:45	A	B	Inder knows about him
0700:47	A		OK 54 seat
0700:50	A	JD	Can I send him now
0700:51	A	B	After about 15–20 minutes
0700:52	A	JD	OK
0700:53	A	B	Thank you
0700:54	A	JD	Welcome Sir
0705:08	A	B	Everybody is having that Whisky and Beer. Whatever you know. That is what I am saying
0705:16	A	B	Without fail – all of them
0705:17	A	B	Whatever is allowed

0705:19	A	SL	Kya karega [what can you do]
0705:20	A	B	Nahin [no]
0705:21	A	SL	It is a hard core problem
0705:22	A	B	Nahin [no]
0705:24	A	B	All I mean, all girls, all ———
0705:30	A		——— Cabin Crew
0705:34	A	B	Cabin crew. Everybody
0705:58	A	N	They must be carrying for somebody else also
0706:00	A	B	Somebody Ten beers. Six beers
0706:03	A	N	Ten beers
0706:05	A		May be some — — —
0706:07	A		Hold. Hold
0706:39	A, C	AI-182(B)	Shanwick AI-182. Good morning
0706:49	A, C	Shannon	Station calling Shannon Go ahead again
0706:54	A, C	AI-182(B)	AI-182. AI-182 is 51 N 15 W at 0705 level 310. Estimate FIR 08 W 51 N 08 W at 0735
0707:10	A, C	Shannon	182 your correct Shannon frequency is 131 15
0707:16	A, C	AI-182(B)	131 15 Sir
0707:22	A, C	TWA 770	——— Merley and upper red 37 to Ibsley and the standard routing
0707:32	A, C	Shannon	770 that checks. Maintain 350
0707:34	A, C	TWA 770	Maintain 35 roger
0707:39	A, C	Shannon	Empress 282 Squawk 2011 and go ahead your progress please
0707:43	A, C	CP 282	282 is 51 N 15 W at 07 370 estimating 51 N 08 W at 35
0707:53	A, C	Shannon	Empress 282 roger cleared Amsterdam via 51 N 08 W upper blue 40 to Brecon upper green 1 Woodley upper blue 29 to Clacton and upper red 1 North to Amsterdam
0708:10	A, C	CP 282	Empress 282 is cleared to Amsterdam 51 N 08 W upper blue 40 Brecon green 1 Woodley Clacton and upper red 1 Amsterdam

0708:21	A, C	Shannon	Empress 282 that checks. Maintain flight level 370
0708:25	A, C	CP 282	Roger 370
0708:28	A, C	AI-182(B)	AI-182 Good morning
0708:34	A, C	Shannon	AI-182 Good morning Squawk 2005 and go ahead please
0708:38	A, C	AI-182(B)	3005 Squawking and AI-182 is 51 N 15 W at 0705 level 310 estimate FIR 51 N 08 W at 0735 and BUNTY next
0709:02	A, C	Shannon	AI-182 Shannon Roger cleared London VIA 51 N 08 W BUNTY upper blue 40 to Merley – upper Red 37 to Ibsley F 310
0709:19	A, C	AI-182(B)	AI-182 is cleared to London Via Upper Bravo 40 to BUNTY then Upper Red 37 to Ibsley and maintain 310
0709:37	A, C	Shannon	AI-182 after BUNTY maintain the upper Blue 40 to Merley and then upper Red 37
0709:44	A, C	AI-182(B)	Right air BUNTY to Merley and then upper Red 37 AI-182
0709:49	A, C	Shannon	AI-182 would you squawk 2005 I repeat 2005
0709:58	A, C	AI-182(B)	Right sir Squawking 2005 182
	A		——
	A		——
0711:24	A		Hann bhai bulao [OK brother]
	A		——
0711:30	A		Kaun sa form hai [what forms are those?]
0711:38	A		That is all right
	A		———
	A		——— (Feeble words)
0713:42	A	D	Bhinder
0713:43	A	B	Haan Ji [yes]
0713:44	A	D	London operation ko bolna [tell London operations]
0713:45	A	B	Hun

0713:47	A	D	Flight Purser. They want about 30 Custom seals
0713:50	A	B	Customs
0713:52	A	D	Custom seals. Wo Bar seal karane ke liye [to seal the bar]
0713:54	A		For their arrival – Customs. Bar———
0714:01	A, C, P, F		(Sound) No further recording

Channel Identification
A — Area Mike
C — Co-pilot's Channel
P — Pilot's Channel
F — Flight Engineer's Channel

Voice Identification
N — Capt Narendra (Commander)
B — Capt Bhinder (Co-pilot)
D — Mr Dumasia (Flight Engineer)
JD — J Dinshaw (Att Flight Purser)
SL — Sampath Lazar (In Flight Supervisor)

Note: There is no recording on Pilot's and Flight Engineer's Channels. However, the 'Sound' (Bang) at the end of the tape has appeared on all the channels i.e. on Pilot's Co-pilot's, Flight Engineer's and on Area Mike.

3. The Brave Heroes

The noble nature of man often comes to the surface in times of tragedies. And it was exhibited by men of the merchant fleets, Royal Navy vessels and Irish ships as they searched a wide area of the Atlantic off the Kerry Coast on 23 June, 1985. It was a time when men did not care for their own lives as they searched without hope for survivors and at the same time picked up bodies from the ocean.

There was an unexpected peril as the men lowered themselves into the water and came face-to-face with the monsters of the sea. The commanding officer of the Irish warship *Le Aisling*, Lieut. Commander James J. Robinson, gave the following account of the search as his vessel joined others in the area:

'On that morning, we were patrolling off the south-west coast of Ireland. We were roughly 50 miles off shore and at that time we received an alert, at 8:52 local time (7:52 GMT).

'The message from Valentia Radio said that there was an aircraft – they did not specify it at that time – but that it had vanished from the radar screen and its position was 51 north and 12.50 west. On receipt of that message, I immediately headed towards that position.

'At 11:45, we were merely some miles from the scene and we established communication with a merchant vessel, the *Laurentian Forest*. She reported that she was a merchant ship and she was the first ship on the scene. She reported that she had seen wreckage and bodies. The aircraft discharged smoke, lots, identifying the location of the wreckage and the bodies. And we altered course towards the smoke.

'My ship recovered a total of 38 bodies. For my crew to recover bodies it was necessary to lower a Gemini (inflatable boat) into the water and on a number of occasions my men had to go into the water also.

'I had three men in a boat. On each trip . . . a maximum of six bodies were recovered and hoisted aboard or winched on board. They (the three men) had to enter the water on these occasions and they encountered sharks while recovering bodies on a number of occasions.

'And I have recommended three men for awards as a result of that and those men were Able Seaman Brown, Leading Seaman McCarthy and Petty Officer Mahon, who was in charge.'

While some men would rather kill, like the sharks, there are many McCarthys, Browns and Mahons in this world who keep it going, who care more for others than themselves.